Solutions
for
Introductory Chemistry: A Foundation
Introductory Chemistry
Basic Chemistry

Third Editions

Zumdahl

James F. Hall
University of Massachusetts at Lowell

D. C. Heath and Company

Lexington, Massachusetts   Toronto

Copyright © 1996 by D. C. Heath and Company.

Previous editions copyright © 1990, 1993 by D. C. Heath and Company.

All rights reserved. No portion of this publication may be reproduced or transmitted in any form or by any means, electronic or mechanical, including photocopy, recording, or any information storage or retrieval system, without permission in writing from the publisher.

Published simultaneously in Canada.

Printed in the United States of America.

International Standard Book Number: 0-669-39958-2

10 9 8 7 6 5 4 3 2 1

Table of Contents

Preface..iv
Chapter 1   Chemistry: An Introduction..................................1
Chapter 2   Measurements and Calculations...............................2
Chapter 3   Matter and Energy..15
Cumulative Review: Chapters 1, 2, and 3................................22
Chapter 4   Chemical Foundations: Elements, Atoms, and Ions...........27
Chapter 5   Nomenclature...35
Cumulative Review: Chapters 4 and 5....................................43
Chapter 6   Chemical Reactions: An Introduction........................48
Chapter 7   Reactions in Aqueous Solutions.............................56
Chapter 8   Classifying Chemical Reactions.............................63
Cumulative Review: Chapters 6, 7, and 8................................69
Chapter 9   Chemical Composition.......................................73
Chapter 10   Chemical Quantities.......................................101
Cumulative Review: Chapters 9 and 10..................................123
Chapter 11   Modern Atomic Theory......................................130
Chapter 12   Chemical Bonding..137
Cumulative Review: Chapters 11 and 12.................................151
Chapter 13   Gases...160
Chapter 14   Liquids and Solids..177
Chapter 15   Solutions...183
Cumulative Review: Chapters 13, 14, and 15............................200
Chapter 16   Equilibrium...207
Chapter 17   Acids and Bases...216
Cumulative Review: Chapters 16 and 17..................................225
Chapter 18   Oxidation-Reduction Reactions/Electrochemistry............229
Chapter 19   Radioactivity and Nuclear Energy..........................242
Chapter 20   Organic Chemistry...248
Chapter 21   Biochemistry..261

Preface

This guide contains detailed solutions for the even-numbered end-of-chapter problems in the third editions of *Introductory Chemistry*, *Introductory Chemistry: A Foundation*, and *Basic Chemistry* by Steven S. Zumdahl. Hundreds of new problems and questions have been prepared for the third editions of the text, including entirely new *Cumulative Review* exercises, which we hope will be of even greater help to you in gaining an understanding of the fundamental principles of chemistry.

Think of this guide as the last chapter of a mystery novel. Just as you wouldn't initially turn to the last few pages of a mystery to find out "whodunit," you should not look at the solutions in this guide until you have tried diligently to solve the end-of-chapter problems yourself. To learn chemistry effectively, you have to spend a lot of time pushing a pencil around on paper!

As you study chemistry, first go over your lecture notes and read through your textbook to see if there is any factual material or terminology you don't understand. Then work through the example problems your professor had done in class and the example problems and self-check exercises in the textbook, paying close attention to the *method* of solution used for each problem. Then try again to work through the *same* classroom and textbook examples yourself *without* looking at the solutions. Once you feel comfortable with the example problems, you can feel confident to tackle the end-of-chapter problems. After you've mastered a few chapters, then test your understanding by going through the *Cumulative Review* material provided for those chapters. These reviews tie together the material covered in the previous few chapters and are intended to give you some perspective on your progress in chemistry.

One topic that causes students much concern is the matter of significant figures, and the determination of the number of digits to which a solution to a problem should be reported. To avoid truncation errors in the solutions contained in this guide, the solutions typically report *intermediate* answers to one more digit that appropriate for the final answer. The *final* answer to each problem is then given to the correct number of significant figures based on the data provided in the problem. Realize especially that your calculator doesn't know anything about the rules for significant figures: a calculator will report answers to however many digits it is programmed for, regardless of whether those digits mean anything. *You* have to have the understanding of the measurements involved and have to do the rounding of answers yourself.

Many very dedicated people have worked long hours to prepare this guide. Particular thanks go to John Goodenow of Lawrence Technological University and John Keetch of Southwestern College for their assistance in checking for errors in the manuscript, and to June Goldstein of D. C. Heath for her patience and good cheer.

Good luck in your study of chemistry!

James F. Hall

Chapter 1   Chemistry: An Introduction

2. Examples: physician (understanding cellular processes, understanding how drugs and bloodtests work); farmer (understanding which pesticide or fertilizer is needed and how these chemicals work); photographer (understanding how the film exposure and developing chemical processes occur and how they can be controlled and modified).

4. There are, unfortunately, many examples. Chemical and biological weapons are still being produced in some countries. Although the development of new plastics has been a boon in many endeavors, this also increases our depletion of fossil fuels and our solid waste problems. Although many exciting new drugs and treatments have become available, the same biotechnology may lead to testing procedures for determining whether a person has a genetic likelihood of developing a particular disease, which may make it impossible or difficult for that person to obtain health or life insurance.

6. This answer depends on your own experience, but consider this example. I got a "ding" in the side of my car a few months ago. I didn't think the paint had been badly damaged, but apparently it had. This part of the car is made of steel (basically iron), and I'm beginning to see a rust (iron oxide) stain appear where the ding is. Here I have mentioned a chemical substance (iron) and have described how I know the iron is chemically changing.

8. Answer will depend on student experience.

10.
    a. quantitative - a number (measurement) is indicated explicitly
    b. qualitative - only a qualitative description is given
    c. quantitative - a numerical measurement is indicated
    d. qualitative - only a qualitative description is given
    e. quantitative - a number (measurement) is implied
    f. qualitative - a qualitative judgment is given
    g. quantitative - a numerical quantity is indicated

12. A natural law is a *summary of observed, measurable behavior* that occurs repeatedly and consistently. A theory is our attempt to *explain* such behavior.

14. Most applications of chemistry are oriented toward the interpretation of observations and the solving of problems. Although memorization of some facts may *aid* in these endeavors, it is the ability to combine, relate, and synthesize information that is most important in the study of chemistry.

16. In real life situations, the problems and applications likely to be encountered are not simple textbook examples. One must be able to observe an event, hypothesize a cause, and then test this hypothesis. One must be able to carry what has been learned in class forward to new, different situations.

Chapter 2   Measurements and Calculations

2. The decimal point must be moved three places, giving three as the power of ten needed.

4. Because 0.0021 is less than one, the exponent will be *negative*. Because 4540 is greater than one, the exponent will be *positive*.

6.
   a. −5; 6.7×10^{-5}

   b. 6; 9.331442×10^{6}

   c. −4; 1×10^{-4}

   d. 4; 1.631×10^{4}

8.
   a. The decimal point must be moved two places to the right, so the exponent is negative 2; $0.04731 = 4.731 \times 10^{-2}$

   b. The decimal point must be moved three places to the left, so the exponent is positive 3; $4284 = 4.284 \times 10^{3}$

   c. The decimal point does not have to be moved, so the exponent is zero; $4.201 = 4.201 \times 10^{0}$

   d. The decimal point must be moved ten places to the right, so the exponent is negative 10; $0.000000000141 = 1.41 \times 10^{-10}$

   e. The decimal point must be moved one place to the left, so the exponent is positive 1; $52.3 = 5.23 \times 10^{1}$

   f. The decimal point must be moved two places to the right, so the exponent is negative 2; $0.04909 = 4.909 \times 10^{-2}$

   g. The decimal point must be moved seven places to the left, so the exponent is positive seven; $54,331,000 = 5.4331 \times 10^{7}$

   h. The decimal point must be moved one place to the right, so the exponent is negative 1; $0.981 = 9.81 \times 10^{-1}$

10.
    a. The decimal point must be moved two places to the right; $4.83 \times 10^{2} = 483$

    b. The decimal point must be moved four places to the left; $7.221 \times 10^{-4} = 0.0007221$

    c. The decimal point does not have to be moved; $6.1 \times 10^{0} = 6.1$

    d. The decimal point must be moved eight places to the left; $9.11 \times 10^{-8} = 0.0000000911$

    e. The decimal point must be moved six places to the right; $4.221 \times 10^{6} = 4,221,000$

f. The decimal point must be moved three places to the left; $1.22 \times 10^{-3} = 0.00122$

g. The decimal point must be moved three places to the right; $9.999 \times 10^3 = 9,999$

h. The decimal point must be moved five places to the left; $1.016 \times 10^{-5} = 0.00001016$

i. The decimal point must be moved five places to the right; $1.016 \times 10^5 = 101,600$

j. The decimal point must be moved one place to the left; $4.11 \times 10^{-1} = 0.411$

k. The decimal point must be moved four places to the right; $9.71 \times 10^4 = 97,100$

l. The decimal point must be moved four places to the left; $9.71 \times 10^{-4} = 0.000971$

12. To say that scientific notation is in *standard* form means that you have a number between 1 and 10, followed by an exponential term. The numbers given in this problem are *not* between 1 and 10 as written.

   a. $131.2 \times 10^{-3} = (1.312 \times 10^2) \times 10^{-3} = 1.312 \times 10^{-1}$

   b. $14.72 \times 10^2 = (1.472 \times 10^1) \times 10^2 = 1.472 \times 10^3$

   c. $1,201 \times 10^{-6} = (1.201 \times 10^3) \times 10^{-6} = 1.201 \times 10^{-3}$

   d. $44.3 \times 10^4 = (4.43 \times 10^1) \times 10^4 = 4.43 \times 10^5$

   e. $0.00721 \times 10^3 = (7.21 \times 10^{-3}) \times 10^3 = 7.21 \times 10^0 = 7.21$

   f. $0.0914 \times 10^{-4} = (9.14 \times 10^{-2}) \times 10^{-4} = 9.14 \times 10^{-6}$

   g. $0.000129 \times 10^5 = (1.29 \times 10^{-4}) \times 10^5 = 1.29 \times 10^1$

   h. $0.00001901 \times 10^{-6} = (1.901 \times 10^{-5}) \times 10^{-6} = 1.901 \times 10^{-11}$

14. a. $1/0.00032 = 3.1 \times 10^3$

   b. $10^3/10^{-3} = 1 \times 10^6$

   c. $10^3/10^3 = 1$ (1×10^0); any number divided by itself is unity.

   d. $1/55,000 = 1.8 \times 10^{-5}$

   e. $(10^5)(10^4)(10^{-4})/10^{-2} = 1 \times 10^7$

4   Chapter 2   Measurements and Calculations

    f. $43.2/(4.32 \times 10^{-5}) = \dfrac{4.32 \times 10^{1}}{4.32 \times 10^{-5}} = 1.00 \times 10^{6}$

    g. $(432 \times 10^{-5})/432 = \dfrac{4.32 \times 10^{-5}}{4.32 \times 10^{2}} = 1.00 \times 10^{-7}$

    h. $1/(10^{5})(10^{-6}) = 1/(10^{-1}) = 1 \times 10^{1}$

16. grams

18.
   a. mega-
   b. milli-
   c. nano-
   d. mega-
   e. centi-
   f. micro-

20. A mile represents, by definition, a greater distance than a kilometer. Therefore 100 mi represents a greater distance than 100 km.

22. quart

24. kilogram

26. 1.75 m is approximately 5 ft, 9". Therefore the man is slightly taller.

28. d

30. d (the other units would give very large numbers for the distance)

32. Table 2.6 indicates that the diameter of a quarter is 2.5 cm.

$$1 \text{ m} \times \dfrac{100 \text{ cm}}{1 \text{ m}} \times \dfrac{1 \text{ quarter}}{2.5 \text{ cm}} = 40 \text{ quarters}$$

34. uncertainty

36. The scale of the ruler shown is only marked to the nearest *tenth* of a centimeter; writing 2.850 would imply that the scale was marked to the nearest *hundredth* of a centimeter (and that the zero in the thousandths place had been estimated).

38.
   a. infinite (a definition)
   b. one
   c. infinite (a definition)
   d. one
   e. infinite (a definition)

40. final

42. a. 3.13×10^2
    b. 3.13×10^{-4}
    c. 3.13×10^7
    d. 3.13×10^{-1}
    e. 3.13×10^{-2}

44. a. 3.42×10^{-4}
    b. 1.034×10^4
    c. 1.7992×10^1
    d. 3.37×10^5

46. decimal

48. three

50. none

52. a. 641.0 (the answer can only be given to one decimal place, since 212.7 and 26.7 are only given to one decimal place)

    b. 1.327 (the answer can only be given to three decimal places, since 0.221 is only given to three decimal places)

    c. 77.34 (the answer can only be given to two decimal places, since 26.01 is only given to two decimal places)

    d. Before performing the calculation, the numbers have to be converted so that they contain the same power of ten.
    $2.01 \times 10^2 + 3.014 \times 10^3 = 2.01 \times 10^2 + 30.14 \times 10^2 = 32.15 \times 10^2$
    This answer should then be converted to *standard* scientific notation, $32.15 \times 10^2 = 3.215 \times 10^3 = 3,215$.

54. a. 124 (the answer can only be given to three significant figures because 0.995 is only given to three significant figures)

    b. 1.995×10^{-23} (the answer can only be given to four significant figures because 6.022×10^{23} is only given to four significant figures)

    c. 1.14×10^{-2} (the answer can only be given to three significant figures because 0.500 is only given to three significant figures. In fact, 0.500 may be ambiguous as to how many significant figures it contains; to make the number of significant figures absolutely clear, 0.500 would be better written as 5.00×10^{-1} to show three significant figures)

6   Chapter 2   Measurements and Calculations

      d. 5.3×10^{-4} (the answer can only be given to two significant figures because 0.15 is only given to two significant figures)

56. a. $(2.0944 + 0.0003233 + 12.22)/7.001 =$

         $(14.314723)/7.001 = 2.045$

    b. $(1.42 \times 10^2 + 1.021 \times 10^3)/(3.1 \times 10^{-1}) =$

         $(142 + 1021)/(3.1 \times 10^{-1}) = (1163)/(3.1 \times 10^{-1}) = 3751 = 3.8 \times 10^3$

    c. $(9.762 \times 10^{-3})/(1.43 \times 10^2 + 4.51 \times 10^1) =$

         $(9.762 \times 10^{-3})/(143 + 45.1) = (9.762 \times 10^{-3})/(188.1) = 5.19 \times 10^{-5}$

    d. $(6.1982 \times 10^{-4})^2 = (6.1982 \times 10^{-4})(6.1982 \times 10^{-4}) = 3.8418 \times 10^{-7}$

58. an infinite number (a definition)

60. $\dfrac{1 \text{ ft}}{12 \text{ in.}}$ $\dfrac{12 \text{ in.}}{1 \text{ ft}}$

62. $\dfrac{1 \text{ lb}}{\$0.79}$

64. a. $2.23 \text{ m} \times \dfrac{1.094 \text{ yd}}{1 \text{ m}} = 2.44 \text{ yd}$

    b. $46.2 \text{ yd} \times \dfrac{1 \text{ m}}{1.094 \text{ yd}} = 42.2 \text{ m}$

    c. $292 \text{ cm} \times \dfrac{1 \text{ in.}}{2.54 \text{ cm}} = 115 \text{ in.}$

    d. $881.2 \text{ in.} \times \dfrac{2.54 \text{ cm}}{1 \text{ in.}} = 2238 \text{ cm}$

    e. $1043 \text{ km} \times \dfrac{1 \text{ mi}}{1.6093 \text{ km}} = 648.1 \text{ mi}$

    f. $445.5 \text{ mi} \times \dfrac{1.6093 \text{ km}}{1 \text{ mi}} = 716.9 \text{ km}$

    g. $36.2 \text{ m} \times \dfrac{1 \text{ km}}{1000 \text{ m}} = 0.0362 \text{ km}$

    h. $0.501 \text{ km} \times \dfrac{1000 \text{ m}}{1 \text{ km}} \times \dfrac{100 \text{ cm}}{1 \text{ m}} = 5.01 \times 10^4 \text{ cm}$

66. a. $254.3 \text{ g} \times \dfrac{1 \text{ kg}}{1000 \text{ g}} = 0.2543 \text{ kg}$

Chapter 2   Measurements and Calculations   7

b. $2.74 \text{ kg} \times \dfrac{1000 \text{ g}}{1 \text{ kg}} = 2.74 \times 10^3 \text{ g}$

c. $2.74 \text{ kg} \times \dfrac{1 \text{ lb}}{0.45359 \text{ kg}} = 6.04 \text{ lb}$

d. $2.74 \text{ kg} \times \dfrac{1 \text{ lb}}{0.45359 \text{ kg}} \times \dfrac{16 \text{ oz}}{1 \text{ lb}} = 96.7 \text{ oz}$

e. $534.1 \text{ g} \times \dfrac{1 \text{ lb}}{453.59 \text{ g}} = 1.177 \text{ lb}$

f. $1.75 \text{ lb} \times \dfrac{453.59 \text{ g}}{1 \text{ lb}} = 794 \text{ g}$

g. $8.7 \text{ oz} \times \dfrac{1 \text{ lb}}{16 \text{ oz}} \times \dfrac{453.59 \text{ g}}{1 \text{ lb}} = 2.5 \times 10^2 \text{ g}$

h. $45.9 \text{ g} \times \dfrac{1 \text{ lb}}{453.59 \text{ g}} \times \dfrac{16 \text{ oz}}{1 \text{ lb}} = 1.62 \text{ oz}$

68. 190 mi = 1.9×10^2 mi to two significant figures

$1.9 \times 10^2 \text{ mi} \times \dfrac{1 \text{ km}}{0.62137 \text{ mi}} = 3.1 \times 10^2 \text{ km}$

$3.1 \times 10^2 \text{ km} \times \dfrac{1000 \text{ m}}{1 \text{ km}} = 3.1 \times 10^5 \text{ m}$

$1.9 \times 10^2 \text{ mi} \times \dfrac{5{,}280 \text{ ft}}{1 \text{ mi}} = 1.0 \times 10^6 \text{ ft}$

70. $1 \times 10^{-10} \text{ m} \times \dfrac{100 \text{ cm}}{1 \text{ m}} = 1 \times 10^{-8} \text{ cm}$

$1 \times 10^{-8} \text{ cm} \times \dfrac{1 \text{ in.}}{2.54 \text{ cm}} = 4 \times 10^{-9} \text{ in.}$

$1 \times 10^{-8} \text{ cm} \times \dfrac{1 \text{ m}}{100 \text{ cm}} \times \dfrac{10^9 \text{ nm}}{1 \text{ m}} = 0.1 \text{ nm}$

72. Celsius

74. 273

76. Fahrenheit (F)

78. $t_C = t_K - 273$

a. $275 - 273 = 2°C$

b. $445 - 273 = 172°C$

c. $0 - 273 = -273°C$

d. $77 - 273 = -196°C$

e. $10{,}000. - 273 = 9727°C$

f. $2.3 - 273 = -271°C$

80. $t_F = 1.80(t_C) + 32$

   a. $1.80(78.1) + 32 = 173 \ °F$

   b. $1.80(40.) + 32 = 104 \ °F$

   c. $1.80(-273) + 32 = -459 \ °F$

   d. $1.80(32) + 32 = 90. \ °F$

82. a. $°C = (°F - 32)/1.80 = (-185 \ °F - 32)/1.80 = (-217)/1.80 = -121 \ °C$

   b. $255 \ K = (255 - 273) \ °C = -18 \ °C$

   $°F = 1.80(-18 \ °C) + 32 = (-32.4) + 32 = 0.4 \ °F$ (approximately 0 °F)

   c. $400. \ °C + 273 = 673 \ K$

   d. $°C = (0 \ °F - 32)/1.80 = -17.8 \ °C$

   $-17.8 \ °C + 273 = 255 \ K$

84. g/cm^3 (g/mL)

86. volume

88. same

90. copper

92. $\text{density} = \dfrac{\text{mass}}{\text{volume}}$

   a. $d = \dfrac{122.4 \ g}{4.3 \ cm^3} = 28 \ g/cm^3$

   b. $v = 0.57 \ m^3 \times \left(\dfrac{100 \ cm}{1 \ m}\right)^3 = 5.7 \times 10^5 \ cm^3$

   $d = \dfrac{1.9302 \times 10^4 \ g}{5.7 \times 10^5 \ cm^3} = 0.034 \ g/cm^3$

c. $m = 0.0175 \text{ kg} \times \dfrac{1000 \text{ g}}{1 \text{ kg}} = 17.5 \text{ g}$

$d = \dfrac{17.5 \text{ g}}{18.2 \text{ mL}} = 0.962 \text{ g/mL} = 0.962 \text{ g cm}^3$

d. $v = 0.12 \text{ m}^3 \times \left(\dfrac{100 \text{ cm}}{1 \text{ m}}\right)^3 = 1.2 \times 10^5 \text{ cm}^3$

$d = \dfrac{2.49 \text{ g}}{1.2 \times 10^5 \text{ cm}^3} = 2.1 \times 10^{-5} \text{ g/cm}^3$

94. $d = \dfrac{55.221 \text{ g}}{67.1 \text{ mL}} = 0.823 \text{ g/mL}$

96. $m = 3.5 \text{ lb} \times \dfrac{453.59 \text{ g}}{1 \text{ lb}} = 1.59 \times 10^3 \text{ g}$

$v = 1.2 \times 10^4 \text{ in}^3 \times \left(\dfrac{2.54 \text{ cm}}{1 \text{ in}}\right)^3 = 1.97 \times 10^5 \text{ cm}^3$

$d = \dfrac{1.59 \times 10^3 \text{ g}}{1.97 \times 10^5 \text{ cm}^3} = 8.1 \times 10^{-3} \text{ g/cm}^3$

The material will float.

98. $5.25 \text{ g} \times \dfrac{1 \text{ cm}^3}{10.5 \text{ g}} = 0.500 \text{ cm}^3 = 0.500 \text{ mL}$

$11.2 \text{ mL} + 0.500 \text{ mL} = 11.7 \text{ mL}$

100. a. $1.00 \times 10^3 \text{ cm}^3 \times \dfrac{11.34 \text{ g}}{1 \text{ cm}^3} = 1.13 \times 10^4 \text{ g}$

$1.00 \text{ m}^3 \times \left(\dfrac{100 \text{ cm}}{1 \text{ m}}\right)^3 \times \dfrac{11.34 \text{ g}}{1 \text{ cm}^3} = 1.13 \times 10^7 \text{ g}$

b. $1.00 \times 10^3 \text{ cm}^3 \times \dfrac{2.16 \text{ g}}{1 \text{ cm}^3} = 2.16 \times 10^3 \text{ g}$

$1.00 \text{ m}^3 \times \left(\dfrac{100 \text{ cm}}{1 \text{ m}}\right)^3 \times \dfrac{2.16 \text{ g}}{1 \text{ cm}^3} = 2.16 \times 10^6 \text{ g}$

c. $1.00 \times 10^3 \text{ cm}^3 \times \dfrac{0.880 \text{ g}}{1 \text{ cm}^3} = 8.80 \times 10^2 \text{ g}$

$1.00 \text{ m}^3 \times \left(\dfrac{100 \text{ cm}}{1 \text{ m}}\right)^3 \times \dfrac{0.880 \text{ g}}{1 \text{ cm}^3} = 8.8 \times 10^5 \text{ g}$

d. $1.00 \times 10^3 \text{ cm}^3 \times \dfrac{7.87 \text{ g}}{1 \text{ cm}^3} = 7.87 \times 10^3 \text{ g}$

10 Chapter 2 Measurements and Calculations

$$1.00 \text{ m}^3 \times \left(\frac{100 \text{ cm}}{1 \text{ m}}\right)^3 \times \frac{7.87 \text{ g}}{1 \text{ cm}^3} = 7.87 \times 10^6 \text{ g}$$

102. a. $3.011 \times 10^{23} = 301,100,000,000,000,000,000,000$
 b. $5.091 \times 10^9 = 5,091,000,000$
 c. $7.2 \times 10^2 = 720$
 d. $1.234 \times 10^5 = 123,400$
 e. $4.32002 \times 10^{-4} = 0.000432002$
 f. $3.001 \times 10^{-2} = 0.03001$
 g. $2.9901 \times 10^{-7} = 0.00000029901$
 h. $4.2 \times 10^{-1} = 0.42$

104. a. centimeters
 b. kilometers
 c. micrometers
 d. millimeters

106. $36.2 \text{ blim} \times \dfrac{1400 \text{ kryll}}{1 \text{ blim}} = 5.07 \times 10^4 \text{ kryll}$

$170 \text{ kryll} \times \dfrac{1 \text{ blim}}{1400 \text{ kryll}} = 0.12 \text{ blim}$

$72.5 \text{ kryll}^2 \times \left(\dfrac{1 \text{ blim}}{1400 \text{ kryll}}\right)^2 = 3.70 \times 10^{-5} \text{ blim}^2$

108. $52 \text{ cm} \times \dfrac{1 \text{ in.}}{2.54 \text{ cm}} = 20. \text{ in.}$

110. $1 \text{ lb} \times \dfrac{1 \text{ kg}}{2.2 \text{ lb}} \times \dfrac{\$1}{F5} \times \dfrac{11.5F}{1 \text{ kg}} = \$1$

112. °X = 1.26°C + 14

114. $d = \dfrac{36.8 \text{ g}}{10.5 \text{ L}} = 3.50 \text{ g/L} \quad (3.50 \times 10^{-3} \text{ g/cm}^3)$

116. for ethanol, $100. \text{ mL} \times \dfrac{0.785 \text{ g}}{1 \text{ mL}} = 78.5 \text{ g}$

 for benzene, $1000 \text{ mL} \times \dfrac{0.880 \text{ g}}{1 \text{ mL}} = 880. \text{ g}$

 total mass, 78.5 + 880. = 959 g

Chapter 2   Measurements and Calculations   11

118. a. negative
 b. zero
 c. positive
 d. negative

120. a. 2; positive
 b. 11; negative
 c. 3; positive
 d. 5; negative
 e. 5; positive
 f. zero; zero
 g. 1; negative
 h. 7; negative

122. a. 2; positive
 b. 3; negative
 c. 1; positive
 d. 8; negative
 e. 4; positive
 f. 1; positive
 g. 6; negative
 h. 1; negative

124. a. The decimal point must be moved three places to the left; $2.98 \times 10^{-5} = 0.0000298$

 b. The decimal point must be moved nine places to the right; $4.358 \times 10^{9} = 4,358,000,000$

 c. The decimal point must be moved six places to the left; $1.9928 \times 10^{-6} = 0.0000019928$

 d. The decimal point must be moved 23 places to the right; $6.02 \times 10^{23} = 602,000,000,000,000,000,000,000$

 e. The decimal point must be moved one place to the left; $1.01 \times 10^{-1} = 0.101$

 f. The decimal point must be moved three places to the left; $7.87 \times 10^{-3} = 0.00787$

g. The decimal point must be moved seven places to the right; $9.87 \times 10^7 = 98,700,000$

h. The decimal point must be moved two places to the right; $3.7899 \times 10^2 = 378.99$

i. The decimal point must be moved one place to the left; $1.093 \times 10^{-1} = 0.1093$

j. The decimal point must be moved zero places; $2.9004 \times 10^0 = 2.9004$

k. The decimal point must be moved four places to the left; $3.9 \times 10^{-4} = 0.00039$

l. The decimal point must be moved eight places to the left; $1.904 \times 10^{-8} = 0.00000001904$

126. a. $1/10^2 = 1 \times 10^{-2}$

b. $1/10^{-2} = 1 \times 10^2$

c. $55/10^3 = \dfrac{5.5 \times 10^1}{1 \times 10^3} = 5.5 \times 10^{-2}$

d. $(3.1 \times 10^6)/10^{-3} = \dfrac{3.1 \times 10^6}{1 \times 10^{-3}} = 3.1 \times 10^9$

e. $(10^6)^{1/2} = 1 \times 10^3$

f. $(10^6)(10^4)/(10^2) = \dfrac{(1 \times 10^6)(1 \times 10^4)}{(1 \times 10^2)} = 1 \times 10^8$

g. $1/0.0034 = \dfrac{1}{3.4 \times 10^{-3}} = 2.9 \times 10^2$

h. $3.453/10^{-4} = \dfrac{3.453}{1 \times 10^{-4}} = 3.453 \times 10^4$

128. kelvin, K

130. centimeter

132. 0.105 m

134. 1 kg (100 g = 0.1 kg)

136. 10 cm (1 cm = 10 mm)

138. 2.8 (the hundredths place is estimated)

Chapter 2   Measurements and Calculations   13

140. a. 0.0000324

b. 7,210,000

c. 2.10×10^{-7}

d. 550,000   (better as 5.50×10^5 to show that the first zero is significant)

e. 200.   (the decimal shows the zeroes are significant)

142. a. 2149.6 (the answer can only be given to the first decimal place, since 149.2 is only known to the first decimal place)

b. 5.37×10^3 (each of the numbers being added is known to the same level of precision). Since the power of ten is the same for each number, the calculation can be performed directly.

c. Before performing the calculation, the numbers have to be converted so that they contain the same power of ten.

$4.03 \times 10^{-2} - 2.044 \times 10^{-3} =$

$4.03 \times 10^{-2} - 0.2044 \times 10^{-2} =$

3.83×10^{-2} (the answer can only be given to three significant figures since 4.03×10^{-2} is only known to three significant figures).

d. Before performing the calculation, the numbers have to be converted so that they contain the same power of ten.

$2.094 \times 10^5 - 1.073 \times 10^6 =$

$2.094 \times 10^5 - 10.73 \times 10^5 =$

-8.64×10^5

144. a. $(2.9932 \times 10^4)(2.4443 \times 10^2 + 1.0032 \times 10^1) =$

$(2.9932 \times 10^4)(24.443 \times 10^1 + 1.0032 \times 10^1) =$

$(2.9932 \times 10^4)(25.446 \times 10^1) = 7.6166 \times 10^6$

b. $(2.34 \times 10^2 + 2.443 \times 10^{-1})/(0.0323) =$

$(2.34 \times 10^2 + 0.002443 \times 10^2)/(0.0323) =$

$(2.34 \times 10^2)/(0.0323) = 7.25 \times 10^3$

c. $(4.38 \times 10^{-3})^2 = 1.92 \times 10^{-5}$

14   Chapter 2   Measurements and Calculations

d. $(5.9938 \times 10^{-6})^{1/2} = 2.4482 \times 10^{-3}$

146. 1 year/12 months; 12 months/1 year

148. a. $908 \text{ oz} \times \dfrac{1 \text{ lb}}{16 \text{ oz}} \times \dfrac{1 \text{ kg}}{2.2046 \text{ lb}} = 25.7 \text{ kg}$

b. $12.8 \text{ L} \times \dfrac{1 \text{ qt}}{0.94633 \text{ L}} \times \dfrac{1 \text{ gal}}{4 \text{ qt}} = 3.38 \text{ gal}$

c. $125 \text{ mL} \times \dfrac{1 \text{ L}}{1000 \text{ mL}} \times \dfrac{1 \text{ qt}}{0.94633 \text{ L}} = 0.132 \text{ qt}$

d. $2.89 \text{ gal} \times \dfrac{4 \text{ qt}}{1 \text{ gal}} \times \dfrac{1 \text{ L}}{1.0567 \text{ qt}} \times \dfrac{1000 \text{ mL}}{1 \text{ L}} = 1.09 \times 10^4 \text{ mL}$

e. $4.48 \text{ lb} \times \dfrac{453.59 \text{ g}}{1 \text{ lb}} = 2.03 \times 10^3 \text{ g}$

f. $550 \text{ mL} \times \dfrac{1 \text{ L}}{1000 \text{ mL}} \times \dfrac{1.0567 \text{ qt}}{1 \text{ L}} = 0.58 \text{ qt}$

150. Assuming exactly 6 gross, 864 pencils

152. a. Celsius temperature = (175 − 32)/1.80 = 79.4 °C

kelvin temperature = 79.4 + 273 = 352 K

b. 255 − 273 = −18 °C

c. (−45 − 32)/1.80 = −43 °C

d. 1.80(125) + 32 = 257 °F

154. $55 \text{ mL} \times \dfrac{0.82 \text{ g}}{1 \text{ mL}} = 45 \text{ g}$

156. $m = 155 \text{ lb} \times \dfrac{453.59 \text{ g}}{1 \text{ lb}} = 7.031 \times 10^4 \text{ g}$

$v = 4.2 \text{ ft}^3 \times \left(\dfrac{12 \text{ in}}{1 \text{ ft}}\right)^3 \times \left(\dfrac{2.54 \text{ cm}}{1 \text{ in}}\right)^3 = 1.189 \times 10^5 \text{ cm}^3$

$d = \dfrac{7.031 \times 10^4 \text{ g}}{1.189 \times 10^5 \text{ cm}^3} = 0.59 \text{ g/cm}^3$

158. a. 23 °F   d. −459 °F
b. 32 °F   e. 187 °F
c. −321 °F   f. −459 °F

Chapter 3   Matter and Energy

2. states

4. container

6. gaseous

8. stronger

10. Gases are easily compressed into smaller volumes, whereas solids and liquids are not. Since a gaseous sample consists mostly of empty space, it is this empty space which is compressed when pressure is applied to a gas.

12. chemical

14. chemical

16. When an electrical current is transferred through water, the current causes the water molecules to break down into their constituent elements (hydrogen gas and oxygen gas). A chemical change occurs whenever one substance is converted into other substances.

18. a. chemical; the patina consists of a compound of copper

   b. chemical; the lye in the drain cleaner chemically breaks down the molecules of protein in the hair

   c. chemical; hydrogen peroxide decomposes into oxygen gas and water when it is applied to a wound (the oxygen gas prevents the growth of anaerobic microorganisms in the wound).

   d. chemical; acids react with the constituent chemicals in teeth producing water soluble products

   e. chemical; marble (which is not soluble in pure water) is chemically converted by acids into water soluble products.

   f. chemical; yeasts chemically convert the sugars in grape juice into alcohol (or vinegar)

   g. physical; the molecules of the substances in mothballs (naphthalene, *p*-dichlorobenzene) still retain their identity when vaporized

   h. physical; the alcohol is just changing its physical state; no new substances are produced

   i. chemical; batteries use oxidation-reduction chemical reactions as a source of electrical current

   j. physical; soap just lowers the surface tension of water, allowing air bubbles to form; the lathering action of soap in

16   Chapter 3   Matter and Energy

        k.   itself is purely a physical process (although the cleansing action of soap may partly be a chemical process)

        k.   chemical; the silver in the spoon chemically reacts with the sulfur in the egg to form black silver sulfide; silver polishes for removing tarnish also make use of a chemical reaction to dissolve away the silver sulfide

20. element

22. elements

24. different from

26. a variable

28. heterogeneous

30. a. although the cotton has undoubtedly been treated with various other substances, cotton basically consists of the carbohydrate cellulose

    b. pure substance

    c. mixture

    d. pure substance

32. a. homogeneous

    b. heterogeneous

    c. heterogeneous

    d. homogeneous

    e. the paper itself is basically homogeneous

34. filtration

36. The solution is heated to vaporize (boil) the water. The water vapor is then cooled so that it condenses back to the liquid state, and the liquid is collected. After all the water is vaporized from the original sample, pure sodium chloride will remain. The process consists of physical changes.

38. the calorie

40. As molecules of liquid water are heated, the heat energy is converted to kinetic energy and the molecules begin to move more quickly and more randomly. Eventually, as heating continues, the kinetic energy of the molecules will be so large that the attractive forces which hold the

molecules together in the volume of the liquid will be overcome. The liquid will then be said to be "boiling," as individual molecules with high kinetic energies separate from the condensed liquid and enter the vapor.

42. temperature

44. Both samples of water are being warmed over the same temperature interval. Since twice as much heat is being applied to the unknown sample as to the 7.40 g sample (2 × 526 J = 1,052 J = 1.052 kJ), then the unknown sample must weigh twice as much (2 × 7.40 g = 14.8 g)

46. Since 1.000 cal = 4.184 J, then 1.000 kcal = 4.184 kJ

    a. $462.4 \text{ kJ} \times \dfrac{1 \text{ kcal}}{4.184 \text{ kJ}} = 110.5 \text{ kcal}$

    b. $18.28 \text{ kJ} \times \dfrac{1 \text{ kcal}}{4.184 \text{ kJ}} = 4.369 \text{ kcal}$

    c. $1.014 \text{ kJ} \times \dfrac{1 \text{ kcal}}{4.184 \text{ kJ}} = 0.2424 \text{ kcal}$

    d. $190.5 \text{ kJ} \times \dfrac{1 \text{ kcal}}{4.184 \text{ kJ}} = 45.53 \text{ kcal}$

48.
    a. $12.30 \text{ kcal} \times \dfrac{1000 \text{ cal}}{1 \text{ kcal}} = 12{,}300 \text{ cal}$

    b. $290.4 \text{ kcal} \times \dfrac{1000 \text{ cal}}{1 \text{ kcal}} = 290{,}400 \text{ cal}$

    c. $940{,}000 \text{ kcal} \times \dfrac{1000 \text{ cal}}{1 \text{ kcal}} = 940{,}000{,}000 \text{ cal}$

    d. $4201 \text{ kcal} \times \dfrac{1000 \text{ cal}}{1 \text{ kcal}} = 4{,}201{,}000 \text{ cal}$

50.
    a. $75.25 \text{ kJ} \times \dfrac{1 \text{ kcal}}{4.184 \text{ kJ}} \times \dfrac{1000 \text{ cal}}{1 \text{ kcal}} = 1.799 \times 10^4 \text{ cal}$

    b. $133{,}200 \text{ cal} \times \dfrac{1 \text{ kcal}}{1000 \text{ cal}} \times \dfrac{4.184 \text{ kJ}}{1 \text{ kcal}} = 557.3 \text{ kJ}$

    c. $232.4 \text{ kcal} \times \dfrac{4.184 \text{ kJ}}{1 \text{ kcal}} \times \dfrac{1000 \text{ J}}{1 \text{ kJ}} = 9.724 \times 10^5 \text{ J}$

    d. $645{,}200 \text{ J} \times \dfrac{1 \text{ cal}}{4.184 \text{ J}} \times \dfrac{1 \text{ kcal}}{1000 \text{ cal}} = 154.2 \text{ kcal}$

18   Chapter 3   Matter and Energy

52. The specific heat capacity of a substance is the heat required to warm one gram of the substance by one Celsius degree.

Temperature increase = 11.2°C

Mass of sample = 1.02 kg = 1.02 × 10³ g

Heat applied to sample = 52.5 kJ = 5.25 × 10⁴ J

$$\text{Specific heat capacity} = \frac{5.25 \times 10^4 \text{ J}}{(11.2°\text{C})(1.02 \times 10^3 \text{ g})} = 4.59 \text{ J/g°C}$$

54. Table 3.2 gives the specific heat of silver as 0.24 J/g °C

Temperature increase = 15.2 −12.0 = 3.2 °C

1.25 kJ = 1250 J

1250 J = (mass of silver) × 0.24 J/g °C × 3.2 °C

mass of silver = 1627 g = 1.6 × 10³ g silver

56. Table 3.2 gives the specific heat capacity of iron as 0.45 J/g °C.

Heat = mass × specific heat capacity × temperature change

4.524 × 10³ J = 742.1 g × 0.45 J/g °C × ΔT

$$\Delta T = \frac{4.524 \times 10^3 \text{ J}}{742.1 \text{ g} \times 0.45 \text{ J/g °C}} = 13.5 = 14 \text{ °C}$$

58. $0.13 \dfrac{\text{J}}{\text{g °C}} \times \dfrac{1 \text{ cal}}{4.184 \text{ J}} = 0.031 \dfrac{\text{cal}}{\text{g °C}}$

$0.13 \dfrac{\text{J}}{\text{g °C}} \times \dfrac{197.0 \text{ g}}{1 \text{ mol}} = 25.6 = 26 \dfrac{\text{J}}{\text{mol °C}}$

60. Specific heat capacities are given in Table 3.2

for gold,   25.0 g × 0.13 J/g °C × 20. °C = 65 J

for mercury, 25.0 g × 0.14 J/g °C × 20. °C = 70. J

for carbon, 25.0 g × 0.71 J/g °C × 20. °C = 360 J

62. 1251 J = 35.2 g × (specific heat capacity) × 25.0 °C

specific heat capacity = 1.42 J/g °C

64. Since X is a pure substance, the fact that two different solids form when electrical current is passed indicates that X must be a compound.

66. Since vaporized water is still the *same substance* as solid water, no chemical reaction has occurred. Sublimation is a physical change.

68. 2.5 kg of water = 2,500 g

    Temperature change = 55.0 − 18.5 = 36.5 °C

    2500 g × 4.184 J/g °C × 36.5 °C = 3.8 × 10⁵ J

70. No calculation is necessary: aluminum will lose more heat because it has the higher specific heat capacity.

72. For any substance, $Q = m \times s \times \Delta T$. The quantity of heat gained by the water in this experiment must equal the total heat lost by the metals (i.e., the *sum* of the amount of heat lost by the iron and the amount of heat lost by the aluminum).

    $(m \times s \times \Delta T)_{water} = (m \times s \times \Delta T)_{iron} + (m \times s \times \Delta T)_{aluminum}$

    If T_f represents the final temperature reached by this system, then

    $[m \times s \times (T_f - 22.5 \,°C)]_{water}$

    $= [m \times s \times (100 \,°C - T_f)]_{iron} + [m \times s \times (100 \,°C - T_f)]_{aluminum}$

    $[97.3 \text{ g} \times 4.184 \text{ J/g °C} \times (T_f - 22.5 \,°C)]$

    $= [10.00 \text{ g} \times 0.24 \text{ J/g °C} \times (100 \,°C - T_f)]$

    $+ [5.00 \text{ g} \times 0.89 \text{ J/g °C} \times (100 \,°C - T_f)]$

    $407.10(T_f - 22.5) = 4.50(100 - T_f) + 4.45(100 - T_f) = 8.95(100 - T_f)$

    $407.10 T_f - 9159.8 = 895 - 8.95 T_f$

    $416.05 T_f = 10,054.8$ which gives $T_f = 24.2 \,°C$

74. $(m \times s \times \Delta T)_{water} = (m \times s \times \Delta T)_{iron}$

    Let T_f represent the final temperature reached by the system

    $[75 \text{ g} \times 4.184 \text{ J/g °C} \times (T_f - 20.)] = [25.0 \text{ g} \times 0.45 \text{ J/g °C} \times (85 - T_f)]$

    $313.8 \,(T_f - 20.) = 11.25 \,(85 - T_f)$

    $313.8 T_f - 6276 = 956.25 - 11.25 T_f$

    $325.05 T_f = 7232.25$

    $T_f = 22.2 \,°C = 22 \,°C$

76. far apart

20   Chapter 3   Matter and Energy

78. chemical

80. chemical

82. electrolysis

84. a. mixture

b. pure substance

c. mixture

d. mixture

e. mixture

86. 9.0 J (It requires twice as much heat to warm twice as large a sample over the same temperature interval.)

88. a. $150.0 \text{ cal} \times \dfrac{4.184 \text{ J}}{1 \text{ cal}} = 627.6 \text{ J}$

b. $4462 \text{ cal} \times \dfrac{4.184 \text{ J}}{1 \text{ cal}} = 1.867 \times 10^4 \text{ J}$

c. $10.0 \text{ cal} \times \dfrac{4.184 \text{ J}}{1 \text{ cal}} = 41.8 \text{ J}$

d. $4.184 \text{ cal} \times \dfrac{4.184 \text{ J}}{1 \text{ cal}} = 17.51 \text{ J}$

90. a. 12.5 kJ

b. 2.99 kcal

c. 2.143 kJ

d. 3.39 kcal

92. The specific heat capacity of water is 4.184 J/g °C.

Temperature increase = 39 − 25 = 14 °C.

75 g × 4.184 J/g °C × 14 °C = 4400 J (to 2 significant figures)

94. For any substance, $Q = m \times s \times \Delta T$. The basic calculation for each of the substances is the same, differing only in the specific heat capacity of the substance in question (specific heat capacities are found in Table 3.2)

Heat required = 150. g × (specific heat capacity) × 11.2 °C

Substance	Specific Heat Capacity	Heat Required
water (l)	4.184 J/g °C	7.03×10^3 J
water (s)	2.03 J/g °C	3.41×10^3 J
water (g)	2.0 J/g °C	3.4×10^3 J
aluminum	0.89 J/g °C	1.5×10^3 J
iron	0.45 J/g °C	7.6×10^2 J
mercury	0.14 J/g °C	2.4×10^2 J
carbon	0.71 J/g °C	1.2×10^3 J
silver	0.24 J/g °C	4.0×10^2 J
gold	0.13 J/g °C	2.2×10^2 J

96. Heat = mass × specific heat capacity × temperature change

    The temperature change is 118 − 25 = 93 °C

    675 J = 55.0 g × s × 93 °C

    $$s = \frac{675 \text{ J}}{55.0 \text{ g} \times 93 \text{ °C}} = 0.13 \text{ J/g °C}$$

Cumulative Review: Chapters 1, 2, and 3

2. By now, after having covered three chapters in this book, it is hoped that you have adopted an "active" approach to your study of chemistry. You may have discovered (perhaps through a disappointing grade on a quiz (though we hope not), that you really have to get involved with chemistry. You can't just sit and take notes, or just look over the solved examples in the textbook. You have to learn to solve problems. You have to learn how to interpret problems, and how to reduce them to the simple mathematical relationships you have studied. Whereas in some courses you might get by on just giving back on exams the facts or ideas presented in class, in chemistry you have to be able to extend and synthesize what has been discussed, and to apply the material to new situations. Don't get discouraged if this is difficult at first: it's difficult for everyone at first.

4. It is difficult sometimes for students (especially beginning students) to understand why certain subjects are required for a given college major. The faculty of your major department, however, have collectively many years of experience in the subject in which you have chosen to specialize. They really do know what courses will be helpful to you in the future. They may have had trouble with the same courses that now give you trouble, but they realize that all the work will be worth it in the end. Some courses you take, particularly in your major field itself, have obvious and immediate utility. Other courses, oftentimes chemistry included, are provided to give you a general background knowledge, which may prove useful in understanding your own major, or other subjects related to your major. In perhaps a burst of bravado, chemistry has been called "the central science" by one team of textbook authors. This moniker is very true however: in order to understand biology, physics, nutrition, farming, home economics, or whatever(it helps to have a general background in chemistry.

6. Whenever a scientific measurement is made, we always employ the instrument or measuring device we are using to the limits of its precision. On a practical basis, this usually means that we *estimate* our reading of the last significant figure of the measurement. An example of the uncertainty in the last significant figure is given for measuring the length of a pin in the text in Figure 2.5. Scientists appreciate the limits of experimental techniques and instruments, and always assume that the last digit in a number representing a measurement has been estimated. Since the last significant figure in every measurement is assumed to be estimated, it is never possible to exclude uncertainty from measurements. The best we can do is to try to improve our techniques and instruments so that we get more significant figures for our measurements.

8. Dimensional analysis is a method of problem solving which pays particular attention to the units of measurements, and uses these units as if they were algebraic symbols that multiply, divide, and cancel. Consider the following example. A dozen of eggs costs $1.25. Suppose we want to know how much one egg costs, and also how much three dozens of eggs will cost. To solve these problems, we need to make use of two equivalence statements:

1 dozen of eggs = 12 eggs

1 dozen of eggs = $1.25

The first of these equivalence statements is obvious: everyone knows that 12 eggs is "equivalent" to one dozen. The second statement also expresses an equivalence, however: if you give the grocer $1.25, he or she will give you a dozen of eggs. From these equivalence statements, we can construct the conversion factors we need to answer the two questions. For the first question (what does one egg cost) we can set up the calculation as follows

$$\frac{\$1.25}{12 \text{ eggs}} = \$0.104 = \$0.10$$

as the cost of one egg   Similarly, for the second question (what will 3 dozens of eggs cost), we can set up the conversion as follows

$$3 \text{ dozens} \times \frac{\$1.25}{1 \text{ dozen}} = \$3.75$$

as the cost of three dozens of eggs. See section 2.6 of the text for how we construct conversion factors from equivalence statements.

10. Defining what scientists mean by "matter" often seems circular to students. Scientists say that matter is something that "has mass and occupies space", without ever really explaining what it means to "have mass" or to "occupy space"! The concept of matter is so basic and fundamental, that it becomes difficult to give a good textbook definition, other than to say that matter is the "stuff" of which everything is made. Matter can be classified and subdivided in many ways, depending on what we are trying to demonstrate.
    On the most fundamental basis, all matter is composed of tiny particles (such as protons, electrons, neutrons, and the other subatomic particles). On one higher level, these tiny particles are combined in a systematic manner into units called atoms. Atoms in turn may be combined to constitute molecules. And finally, large groups of molecules may be placed together to form a bulk sample of substance that we can see.
    Matter can also be classified as to the physical state a particular substance happens to take. Some substances are solids, some a liquids, and some are gases. Matter can also be classified as to whether it is a pure substance (one type of molecule) or a mixture (more than one type of molecule), and furthermore whether a mixture is homogeneous or heterogeneous.

12. Chemists tend to give a functional definition of what they mean by an "element": an element is a fundamental substance that cannot be broken down into any simpler substances by chemical methods. Compounds, on the other hand, can be broken down into simpler substances (often the elements of which the compound is composed). For example, sulfur and oxygen are both elements (sulfur occurs as S_8 molecules and oxygen as O_2 molecules). When sulfur and oxygen are placed together and heated, the

compound sulfur dioxide (SO_2) forms. When we analyze the sulfur dioxide produced, we notice that each and every molecule consists of one sulfur atom and two oxygen atoms, and on a mass basis, consists of 50% each of sulfur and oxygen. We describe this by saying that sulfur dioxide has a constant composition. The fact that a given compound has constant composition is usually expressed in terms of the mass percentages of the elements present in the compound, but realize that the reason the mass percentages are constant is because of a constant number of atoms of each type present in the compound's molecules. If a scientist anywhere in the universe analyzed sulfur dioxide, he or she would find the same composition: if a scientist finds something that does not have the same composition, then the substance cannot be sulfur dioxide.

14. Scientists define energy as "the capacity to do work". As with trying to define "matter" earlier in this chapter, energy is such a fundamental concept that it is hard to define (what is "work"?). Although the *SI* unit of energy is the *joule*, until relatively recently, energies were more commonly given in terms of the *calorie*: one calorie is defined to be the amount of heat required to raise the temperature of one gram of water by one Celsius degree. The calorie is a "working" definition, and we can more easily appreciate this amount of energy. In terms of the *SI* unit, 1 calorie = 4.184 joule, so it takes 4.184 J to raise the temperature of one gram of water by one Celsius degree.

 The specific heat capacity of a substance, in general, is the amount of energy required to raise the temperature of one gram of a substance by one Celsius degree. Therefore, the specific heat capacity of water must be 1.000 cal/g°C, or 4.184 J/g°C. To see how specific heat capacities may be used to calculate the energy change for a process, consider this example: How much energy is required to warm 25.0 g of water from 15.1°C to 35.2°C? The specific heat capacity of water is 1.000 cal/g°C: this is the quantity of energy required to raise the temperature of only *one* gram of water by only *one* Celsius degree. In this example, we are raising the temperature of 25.0 g of water, and we are raising the temperature by (35.2 - 15.1) = 20.1°C. So, using the specific heat capacity as a conversion factor, we can say

$$\text{energy required} = \left(\frac{1.00 \text{ cal}}{\text{g} \, °\text{C}}\right) \times (25.0 \text{ g}) \times (20.1°\text{C}) = 503 \text{ cal}$$

Notice how the units of g and °C cancel, leaving the answer in energy units only. This sort of calculation of energy change can be done for any substance, using the substance's own specific heat capacity (see Table 3.2).

16. a. $7.50 \text{ mi} \times \dfrac{1.6093 \text{ km}}{1 \text{ mi}} = 12.1 \text{ km}$

 b. $6.91 \text{ kg} \times \dfrac{2.2046 \text{ lb}}{1 \text{ kg}} = 15.2 \text{ lb}$

c. $5{,}550 \text{ mL} \times \dfrac{1 \text{ L}}{1000 \text{ mL}} = 5.55 \text{ L}$

d. $2.10 \text{ L} \times \dfrac{1.0567 \text{ qt}}{1 \text{ L}} = 2.22 \text{ qt}$

e. $2.10 \text{ L} \times \dfrac{1 \text{ gal}}{3.7854 \text{ L}} = 0.555 \text{ gal}$

f. $1.315 \text{ m} \times \dfrac{1000 \text{ mm}}{1 \text{ m}} = 1{,}315 \text{ mm}$

g. $12.5 \text{ in} \times \dfrac{1 \text{ yd}}{36 \text{ in}} \times \dfrac{1 \text{ m}}{1.0936 \text{ yd}} = 0.318 \text{ m}$

h. $16 \text{ fl oz} \times \dfrac{0.94633 \text{ L}}{32 \text{ fl oz}} = 0.47 \text{ L}$

i. $2.04 \text{ m} \times \dfrac{1.0936 \text{ yd}}{1 \text{ m}} = 2.23 \text{ yd}$

j. $8.11 \text{ pt} \times \dfrac{3.7854 \text{ L}}{8 \text{ pt}} = 3.84 \text{ L}$

18. $t_F = 1.80(t_C) + 32 \qquad t_C = (t_F - 32)/1.80 \qquad t_K = t_C + 273$

   a. $1.80(-50.1°C) + 32 = -58.2°F$

   b. $(-30.7°C - 32)/1.80 = -34.8°C$

   c. $541 \text{ K} - 273 = 268°C$

   d. $221°C + 273 = 494 \text{ K}$

   e. $351 \text{ K} - 273 = 78°C$

   $1.80(78°C) + 32 = 172.4 = 172°F \; (1.7 \times 10^2 °F$

   f. $(72°F - 32)/1.80 = 22.2°C$

   $22.2°C + 273 = 295.2 = 295 \text{ K}$

20. a. $459 \text{ J} \times \dfrac{1 \text{ cal}}{4.184 \text{ J}} = 109.7 = 110. \text{ cal}$

   b. $7{,}031 \text{ cal} \times \dfrac{4.184 \text{ J}}{1 \text{ cal}} \times \dfrac{1 \text{ kJ}}{1000 \text{ J}} = 29.42 \text{ kJ}$

   c. $55.31 \text{ kJ} = 55{,}310 \text{ J}$

   d. $78.3 \text{ kcal} \times \dfrac{4.184 \text{ kJ}}{1 \text{ kcal}} = 327.6 = 328 \text{ kJ}$

e. 4,541 cal = 4.541 kcal

f. $84.1 \text{ kJ} \times \dfrac{1 \text{ kcal}}{4.184 \text{ kJ}} = 20.1 \text{ kcal}$

Chapter 4   Chemical Foundations: Elements, Atoms and Ions

2. The alchemists discovered several previously unknown elements (mercury, sulfur, antimony) and were the first to prepare several common acids.

4. There are 110 elements presently known; of these 88 occur naturally and 22 are manmade. Table 4.1 lists the most common elements on the earth.

6. The four most abundant elements in living creatures are, respectively, oxygen, carbon, hydrogen, and nitrogen (see Table 4.2). In the nonliving world, the most abundant elements are, respectively, oxygen, silicon, aluminum, and iron (see Table 4.1).

8. Sb (antimony)   Ag (silver)
    Cu (copper)   Na (sodium)
    Au (gold)   Sn (tin)
    Pb (lead)   W (tungsten)
    Hg (mercury)   Fe (iron)
    K (potassium)

10. a. H   d. Ca
    b. Fe   e. Au
    c. Mg   f. He

12. Ir   iridium   Pu   plutonium
    Ta   tantalum   Fr   francium
    Bi   bismuth   At   astatine

14. a. cobalt   e. zinc
    b. silver   f. platinum
    c. chlorine   g. chromium
    d. aluminum   h. sodium

16. According to Dalton, a given compound is always made up of the same number and type of atoms, and so the composition of the compound on a mass percentage basis will always be the same, no matter what the source of the compound is. For example, water molecules from the Atlantic Ocean and the Pacific Ocean all contain two hydrogen atoms bound with one oxygen atom.

18. According to Dalton, all atoms of the same element were *identical*; in particular, every atom of a given element has the same *mass* as every other atom of that element. If a given compound always contains the *same relative numbers* of atoms of each kind, and those atoms always have the *same masses*, then it follows that the compound made from those elements would always contain the same relative masses of its elements.

28 Chapter 4   Chemical Foundations: Elements, Atoms, and Ions

20. a.

b.

c.

d.

e.

f.

Chapter 4 Chemical Foundations: Elements, Atoms, and Ions

22. a. False; Rutherford's bombardment experiments with metal foil suggested that the alpha particles were being deflected by coming near a *dense, positively charged* atomic nucleus.

    b. False; The proton and the electron have opposite charges, but the mass of the electron is *much smaller* than the mass of the proton.

    c. True

24. neutrons

26. neutron; electron

28. electrons

30. False

32. mass

34. Atoms of the same element (i.e., atoms with the same number of protons in the nucleus) may have different numbers of neutrons, and so will have different masses.

36. a. 37   e. 53
    b. 23   f. 95
    c. 46   g. 83
    d. 30   h. 42

38. a. $^{55}_{25}Mn$

    b. $^{25}_{12}Mg$

    c. $^{41}_{20}Ca$

    d. $^{56}_{26}Fe$

    e. $^{238}_{92}U$

    f. $^{244}_{94}Pu$

40. a. 27 protons, 33 neutrons, 27 electrons

    b. 16 protons, 17 neutrons, 16 electrons

c. 4 protons, 6 neutrons, 4 electrons
d. 18 protons, 22 neutrons, 18 electrons
e. 11 protons, 12 neutrons, 11 electrons
f. 36 protons, 48 neutrons, 36 electrons

42.

element	neutrons	atomic number	mass number	symbol
nitrogen	6	7	13	$^{13}_{7}N$
nitrogen	7	7	14	$^{14}_{7}N$
lead	124	82	206	$^{206}_{82}Pb$
iron	31	26	57	$^{57}_{26}Fe$
krypton	48	36	84	$^{84}_{36}Kr$

44. False; The *vertical* columns in the periodic table are referred to as groups or families.

46. Metallic elements are found towards the *left* and *bottom* of the periodic table; there are far more metallic elements than there are nonmetals.

48. hydrogen, nitrogen, oxygen, fluorine, chlorine, plus all the group 8 elements (noble gases)

50. The metalloids are found on either side of the "stairstep" region that is marked on most periodic tables, which divides the metallic elements from the nonmetallic elements. Silicon and germanium are probably the best example of metalloids, since they are used so much in the electronics industry.

52. a. 8; noble gases
     b. 7; halogens
     c. 1; alkali metals
     d. 3
     e. 1; alkali metals
     f. 2; alkaline earth elements
     g. 8; noble gases
     h. 7; halogens

54.

name	symbol	atomic number	group number	metal/nonmetal
rubidium	Rb	37	1	metal
germanium	Ge	32	4	metalloid
magnesium	Mg	12	2	metal
titanium	Ti	22	-	transition metal
iodine	I	53	7	nonmetal

56. Most of the elements are too reactive to be found in the uncombined form in nature, and are found only in compounds.

58. These elements are found *uncombined* in nature and do not readily react with other elements. For many years it was thought that these elements formed no compounds at all, although this has now been shown to be untrue.

60. Diatomic gases: H_2, N_2, O_2, Cl_2, and F_2

    Monatomic gases: He, Ne, Kr, Xe, Rn, and Ar

62. chlorine

64. diamond

66. electrons

68. 2+

70. *-ide*

72. nonmetallic

74.
  a. Fe^{2+}:   26 protons   24 electrons $FeCl_2$
  b. Fe^{3+}:   26 protons   23 electrons $FeCl_3$
  c. Ba^{2+}:   56 protons   54 electrons $BaCl_2$
  d. Cs^+:   55 protons   54 electrons $CsCl$
  e. S^{2-}:   16 protons   18 electrons Na_2S
  f. P^{3-}:   15 protons   18 electrons Na_3P
  g. Br^-:   35 protons   36 electrons $NaBr$
  h. O^{2-}:   8 protons   10 electrons Na_2O

76. a. $2e^-$
   b. $2e^-$
   c. $2e^-$
   d. $3e^-$
   e. $1e^-$
   f. $3e^-$

78. a. Ra^{2+} (element 88, Ra, is in Group 2)
   b. Te^{2-} (element 52, Te, is in Group 6)
   c. I^- (element 53, I, is in Group 7)
   d. Fr^+ (element 87, Fr, is in Group 1)
   e. At^- (element 85, At, is in Group 7)
   f. no ion is likely (element 86, Rn, is a noble gas)

80. Sodium chloride is an *ionic* compound, consisting of Na^+ and Cl^- *ions*. When NaCl is dissolved in water, these ions are *set free*, and can move independently to conduct the electrical current. Sugar crystals, although they may *appear* similar visually contain *no* ions. When sugar is dissolved in water, it dissolves as uncharged *molecules*. There are no electrically charged species present in a sugar solution to carry the electrical current.

82. The total number of positive charges must equal the total number of negative charges so that there will be *no net charge* on the crystals of an ionic compound. A macroscopic sample of compound must ordinarily not have any net charge.

84. a. One 2+ ion is exactly balanced by one 2– ion: BaSe
   b. One 2+ ion is balanced by two 1– ions: CaI_2
   c. Two 1+ ions are needed to balance one 2– ion: K_2S
   d. One 4+ ion is balanced by two 2– ions: SnO_2
   e. One 2+ ion is exactly balanced by one 2– ion: SnO
   f. The smallest whole number that both 3 and 4 divide into is 12; so three 4+ ions will have a total charge of 12+, and for a total charge of 12–, we will need four 3– ions: Pb_3P_4
   g. The smallest whole number that divides both 2 and 3 evenly is 6; so we will need three 2+ ions to balance two 3– ions: Pb_3N_2
   h. One 3+ ion is balanced by three 1– ions: AlI_3

86. a. 6
   b. 1; alkali metals
   c. transition metals

d. 2; alkaline earth elements

e. transition metals

f. transition metals

g. 8; noble gases

h. 2; alkaline earth elements

88.
	element	symbol	atomic number
Group 3	boron	B	5
	aluminum	Al	13
	gallium	Ga	31
	indium	In	49
Group 5	nitrogen	N	7
	phosphorus	P	15
	arsenic	As	33
	antimony	Sb	51
Group 8	helium	He	2
	neon	Ne	10
	argon	Ar	18
	krypton	Kr	36

90. Most of the mass of an atom is concentrated in the nucleus: the *protons* and *neutrons* which constitute the nucleus have similar masses, and these particles are nearly two thousand times heavier than electrons. The chemical properties of an atom depend on the number and location of the *electrons* it possesses. Electrons are found in the outer regions of the atom, and are the particles most likely to be involved in interactions between atoms.

92. $C_6H_{12}O_6$

94. a. 29 protons; 34 neutrons; 29 electrons
   b. 35 protons; 45 neutrons; 35 protons
   c. 12 protons; 12 neutrons; 12 electrons

96. The chief use of gold in ancient times was as *ornamentation*, whether in statuary or in jewelry. Gold possesses an especially beautiful luster, and since it is relatively soft and malleable, it could be worked finely by artisans; among the metals, gold is particularly inert to attack by most substances in the environment.

98. a. Ba   d. B
    b. Br   e. K
    c. Bi   f. P

100. a. Zn
b. U
c. W

102. a. boron
b. radium
c. arsenic
d. calcium
e. mercury
f. bromine
g. fluorine
h. phosphorus

104. a. CO_2
b. $AlCl_3$
c. $HClO_4$
d. SCl_6

106. a. $^{13}_{6}C$
b. $^{13}_{6}C$
c. $^{13}_{6}C$
d. $^{44}_{19}K$
e. $^{41}_{20}Ca$
f. $^{35}_{19}K$

108.

	symbol	number of protons	number of neutrons	mass number
a.	$^{41}_{20}Ca$	20	21	41
b.	$^{55}_{25}Mn$	25	30	55
c.	$^{109}_{47}Ag$	47	62	109
d.	$^{45}_{21}Sc$	21	24	45

Chapter 5   Nomenclature

2. compounds that contain a metal and a nonmetal; compounds between two nonmetals

4. cation

6. sodium ions, Na$^+$, and chloride ions, Cl$^-$

8. Roman numeral

10.
    a. sodium nitride
    b. potassium oxide
    c. silver bromide, silver(I) bromide
    d. magnesium iodide
    e. strontium oxide
    f. calcium hydride
    g. barium chloride
    h. aluminum iodide

12.
    a. Silver's symbol is Ag, not S
    b. correct
    c. Since Na$_3$N is ionic, the prefix *tri-* is not needed.
    d. Since BaCl$_2$ is ionic, the prefix *di* is not needed.
    e. correct

14.
    a. Since the chloride ion has a 1− charge, the lead ion must have a 2+ charge: the name is lead(II) chloride.
    b. Since the oxide ion has a 2− charge, the iron must have a 3+ charge: the name is iron(III) oxide.
    c. Since the iodide ion has a 1− charge, the tin ion must have a 2+ charge: the name is tin(II) iodide.
    d. Since the oxide ion has a 2− charge, the two mercury atoms must each have a 1+ charge: the name is mercury(I) oxide.
    e. Since the sulfide ion has a 2− charge, the mercury ion must have a 2+ charge: the name is mercury(II) sulfide.
    f. Since the iodide ion has a 1− charge, the copper ion must have a 1+ charge: the name is copper(I) iodide.

16.
    a. Since oxide ions have a 2− charge the lead ion must have a 4+ charge: the name is plumb*ic* oxide.

36   Chapter 5   Nomenclature

- b. Since bromide ions have a 1- charge, the tin ion must have a 2+ charge: the name is stann*ous* bromide.

- c. Since the sulfide ion has a 2- charge, the two copper ions must each have a 1+ charge: the name is cupr*ous* sulfide.

- d. Since the iodide ion has a 1- charge, the copper ion must have a 1+ charge: the name is cupr*ous* iodide.

- e. Since iodide ions have a 1- charge, each mercury must have a 1+ charge: the name is mercur*ous* iodide.

- f. Since fluoride ions have a 1- charge, the chromium ion must have a 3+ charge: the name is chrom*ic* fluoride.

18.
- a. germanium tetrahydride
- b. dinitrogen tetrabromide
- c. diphosphorus pentasulfide
- d. selenium dioxide
- e. ammonia (nitrogen trihydride)
- f. silicon dioxide

20.
- a. diboron hexahydride - nonionic (common name: *diborane*)
- b. calcium nitride - ionic
- c. carbon tetrabromide - nonionic
- d. silver(I) sulfide - ionic
- e. copper(II) chloride, cupric chloride - ionic
- f. chlorine monofluoride - nonionic

22.
- a. aluminum oxide - ionic
- b. diboron trioxide - nonionic
- c. dinitrogen tetroxide - nonionic
- d. cobalt(III) sulfide, cobaltic sulfide - ionic
- e. dinitrogen pentoxide - nonionic
- f. aluminum sulfide - ionic

24. oxygen

26. perchlorate, ClO_4^-

28. hypobromite

    IO$_3^-$

    periodate

    OI$^-$

30. a. NO$_3^-$ c. NH$_4^+$
    b. NO$_2^-$ d. CN$^-$

32. a. PO$_4^{3-}$ c. H$_2$PO$_4^-$
    b. P^{3-} d. HPO$_4^-$

34. a. hydrogen carbonate (also commonly called *bi*carbonate)
    b. nitrite
    c. hydrogen sulfate (also commonly called *bi*sulfate)
    d. dihydrogen phosphate
    e. acetate
    f. chlorite

36. a. ammonium nitrate
    b. potassium chlorate
    c. lead(II) sulfate, plumbous sulfate
    d. calcium phosphate
    e. sodium perchlorate
    f. copper(II) hydroxide, cupric hydroxide

38. oxygen (commonly referred to as *oxyacids*)

40. a. hypobromous acid
    b. periodic acid (pronounced *periodic*)
    c. phosphorous acid
    d. chlorous acid
    e. selenic acid
    f. hypoiodous acid
    g. iodous acid
    h. hydrotelluric acid

42. a. PbO_2 d. CuI
    b. $SnBr_2$ e. Hg_2Cl_2
    c. CuS f. CrF_3

44. a. CO_2 d. CI_4
    b. SO_2 e. PF_5
    c. N_2Cl_4 f. P_2O_5

46. a. $Ca_3(PO_4)_2$ d. $BaSO_4$
    b. NH_4NO_3 e. $Fe(NO_3)_3$
    c. $Al(HSO_4)_3$ f. $CuOH$

48. a. HCN e. $HClO$ or $HOCl$
    b. HNO_3 f. HF
    c. H_2SO_4 g. $HBrO_2$
    d. H_3PO_4 h. HF

50. a. Na_2HPO_4 g. $Fe(OH)_3$
    b. $Mg(C_2H_3O_2)_2$ h. SnF_2
    c. SrI_2 i. PbO
    d. H_2Se j. B_2O_3
    e. Ca_3P_2 k. NaH_2PO_4
    f. $AgClO_4$ l. H_3PO_3

52. A moist paste of NaCl would contain Na^+ and Cl^- ions in solution, and would serve as a *conductor* of electrical impulses.

54. $H \rightarrow H^+$ (hydrogen ion: a cation) $+ e^-$

    $H + e^- \rightarrow H^-$ (hydr*ide* ion: an anion)

56. missing oxyanions: IO_3^-; ClO_2^-

    missing oxyacids: $HClO_4$; $HClO$; $HBrO_2$

58. a. gold(III) bromide, auric bromide
    b. cobalt(III) cyanide, cobaltic cyanide
    c. magnesium hydrogen phosphate
    d. diboron hexahydride (diborane is its common name)
    e. ammonia

Chapter 5 Nomenclature 39

    f.   silver(I) sulfate (usually called silver sulfate)

    g.   beryllium hydroxide

60. a.   ammonium carbonate

     b.   ammonium hydrogen carbonate, ammonium bicarbonate

     c.   calcium phosphate

     d.   sulfurous acid

     e.   manganese(IV) oxide

     f.   iodic acid

     g.   potassium hydride

62. a. MCl_4 e. $M(CN)_4$

     b. $M(NO_3)_4$ f. $M(SO_4)_2$

     c. MO_2 g. $M(Cr_2O_7)_2$

     d. $M_3(PO_4)_4$

64. M^+ compounds: MD, M_2E, M_3F

    M^{2+} compounds: MD_2, ME, M_3F_2

    M^{3+} compounds: MD_3, M_2E_3, MF

66.

$CaBr_2$	$Ca(HCO_3)_2$	CaH_2	$Ca(C_2H_3O_2)_2$	$Ca(HSO_4)_2$	$Ca_3(PO_4)_2$
$SrBr_2$	$Sr(HCO_3)_2$	SrH_2	$Sr(C_2H_3O_2)_2$	$Sr(HSO_4)_2$	$Sr_3(PO_4)_2$
NH_4Br	NH_4HCO_3	NH_4H	$NH_4C_2H_3O_2$	NH_4HSO_4	$(NH_4)_3PO_4$
$AlBr_3$	$Al(HCO_3)_3$	AlH_3	$Al(C_2H_3O_2)_3$	$Al(HSO_4)_3$	$AlPO_4$
$FeBr_3$	$Fe(HCO_3)_3$	FeH_3	$Fe(C_2H_3O_2)_3$	$Fe(HSO_4)_3$	$FePO_4$
$NiBr_2$	$Ni(HCO_3)_2$	NiH_2	$Ni(C_2H_3O_2)_2$	$Ni(HSO_4)_2$	$Ni_3(PO_4)_2$
$AgBr$	$AgHCO_3$	AgH	$AgC_2H_3O_2$	$AgHSO_4$	Ag_3PO_4
$AuBr_3$	$Au(HCO_3)_3$	AuH_3	$Au(C_2H_3O_2)_3$	$Au(HSO_4)_3$	$AuPO_4$
KBr	$KHCO_3$	KH	$KC_2H_3O_2$	$KHSO_4$	K_3PO_4
$HgBr_2$	$Hg(HCO_3)_2$	HgH_2	$Hg(C_2H_3O_2)_2$	$Hg(HSO_4)_2$	$Hg_3(PO_4)_2$
$BaBr_2$	$Ba(HCO_3)_2$	BaH_2	$Ba(C_2H_3O_2)_2$	$Ba(HSO_4)_2$	$Ba_3(PO_4)_2$

68. helium

70. iodine

72. 1−

74. 1−

76. a. Mn (manganese), atomic number 25

Mn^{2+} (25 protons, 23 electrons); Mn (25 protons, 25 electrons)

b. Ni (nickel), atomic number 28

Ni^{2+} (28 protons, 26 electrons); Ni (28 protons, 28 electrons)

c. N (nitrogen), atomic number 7

N^{3-} (7 protons, 10 electrons); N (7 protons, 7 electrons)

d. Co (cobalt), atomic number 27

Co^{3+} (27 protons, 24 electrons); Co (27 protons, 27 electrons)

e. Fe (iron), atomic number 26

Fe^{2+} (26 protons, 24 electrons); Fe (26 protons, 26 electrons)

f. P (phosphorus), atomic number 15

P^{3-} (15 protons, 18 electrons); P (15 protons, 15 electrons)

78. a. Two 1+ ions are needed to balance a 2− ion, so the formula must have two Na^+ ions for each S^{2-} ion: Na_2S.

b. One 1+ ion exactly balances a 1− ion, so the formula should have an equal number of K^+ and Cl^- ions: KCl.

c. One 2+ ion exactly balances a 2− ion, so the formula must have an equal number of Ba^{2+} and O^{2-} ions: BaO.

d. One 2+ ion exactly balances a 2− ion, so the formula must have an equal number of Mg^{2+} and Se^{2-} ions: MgSe.

e. One 2+ ion requires two 1− ions to balance charge, so the formula must have twice as many Br^- ions as Cu^{2+} ions: $CuBr_2$.

f. One 3+ ion requires three 1− ions to balance charge, so the formula must have three times as many I^- ions as Al^{3+} ions: AlI_3.

g. Two 3+ ions give a total of 6+, whereas three 2− ions will give a total of 6−. The formula then should contain two Al^{3+} ions and three O^{2-} ions: Al_2O_3.

h. Three 2+ ions are required to balance two 3− ions, so the formula must contain three Ca^{2+} ions for every two N^{3-} ions: Ca_3N_2.

80. a. correct

b. Since AlH$_3$ is ionic, the prefix *tri-* is not needed.

c. correct

d. The hydroxide ion (OH$^-$) has a 1− charge, so iron must be in the 2+ state in this compound; the correct name is iron(II) hydroxide.

e. The chloride ion (Cl$^-$) has a 1− charge, so cobalt must be in the 3+ state in this compound; the correct name is cobalt(III) chloride.

82. a. Since bromide ions always have a 1− charge, the cobalt ion must have a 3+ charge: the name is cobalt*ic* bromide.

b. Since iodide ions always have a 1− charge, the lead ion must have a 4+ charge: the name is plumb*ic* iodide.

c. Since oxide ions always have a 2− charge, and since there are three oxide ions, each iron ion must have a 3+ charge: the name is ferr*ic* oxide.

d. Since sulfide ions always have a 2− charge, the iron ion must have a 2+ charge: the name is ferr*ous* sulfide.

e. Since chloride ions always have a 1− charge, the tin ion must have a 4+ charge: the name is stann*ic* chloride.

f. Since oxide ions always have a 2− charge, the tin ion must have a 2+ charge: the name is stann*ous* oxide.

84. a. iron(III) sulfide, ferric sulfide - ionic

b. gold(III) chloride, auric chloride - ionic

c. arsenic trihydride (common name arsine) - nonionic

d. chlorine monofluoride - nonionic

e. potassium oxide - ionic

f. carbon dioxide - nonionic

86. a. CO$_3^{2-}$

b. HCO$_3^-$

c. C$_2$H$_3$O$_2^-$

d. CN$^-$

88. a. carbonate
    b. chlorate
    c. sulfate
    d. phosphate
    e. perchlorate
    f. permanganate

90. a. $CaCl_2$
    b. Ag_2O
    c. Al_2S_3
    d. $BeBr_2$
    e. H_2S
    f. KH
    g. MgI_2
    h. CsF

92. a. Mg_3P_2
    b. CaF_2
    c. $CoBr_3$
    d. FeI_2
    e. BaO
    f. K_2S

Cumulative Review: Chapters 4 and 5

2. How many elements could you name? While you certainly don't have to memorize all the elements, you should at least be able to give the symbol or name for the most common elements (listed in Table 4.3).

4. Dalton's atomic theory as presented in this text consists of five main postulates. Realize that although Dalton's theory was exceptional scientific thinking for its time, some of the postulates have been modified as our scientific instruments and calculational methods have become increasingly more sophisticated. The main postulates of Dalton's theory are as follows: (1) Elements are made up of tiny particles called atoms; (2) all atoms of a given element are identical; (3) although all atoms of a given element are identical, these atoms are different from the atoms of all other elements; (4) atoms of one element can combine with atoms of another element to form a compound, and such a compound will always have the same relative numbers and types of atoms for its composition; (5) atoms are merely rearranged into new groupings during an ordinary chemical reaction, and no atom is ever destroyed and no new atom is ever created during such a reaction.

6. The expression *nuclear* atom indicates that we view the atom as having a dense center of positive charge (called the nucleus) around which the electrons move through primarily empty space. Rutherford's experiment involved shooting a beam of particles at a thin sheet of metal foil. According to the then current "plum pudding" model of the atom, most of these positively charged particles should have passed right through the foil. However, Rutherford detected that a significant number of particles effectively bounced off something and were deflected backwards to the source of particles, and that other particles were deflected from the foil at large angles. Rutherford realized that his observations could be explained if the atoms of the metal foil had a small, dense, positively charged nucleus, with a significant amount of empty space between nuclei. The empty space between nuclei would allow most of the particles to pass through the atom. However, if an particle hit a nucleus head-on, it would be deflected backwards at the source. If a positively-charged particle passed near a positively charged nucleus (but did not hit the nucleus head-on), then the particle would be deflected by the repulsive forces between the positive charges. Rutherford's experiment conclusively disproved the "plum pudding" model for the atom, which envisioned the atom as a uniform sphere of positive charge, with enough negatively charged electrons scattered through the atom to balance out the positive charge.

8. Isotopes represent atoms of the same element which have different atomic masses. Isotopes are a result of the fact that atoms of a given element may have different numbers of neutrons in their nuclei. Isotopes have the same atomic number (number of protons in the nucleus) but have different mass numbers (total number of protons and neutrons in the nucleus). The different isotopes of an atom are indicated by symbolism of the form $^{A}_{Z}X$ in which Z represents the atomic number, and A the mass number, of element X. For example, $^{13}_{6}C$ represents a nuclide of carbon with atomic number 6 (6 protons in the nucleus) and mass number 13 (reflecting 6 protons plus 7 neutrons in the nucleus). The various isotopes of an element have identical chemical properties since the

chemical properties of an atom are a function of the electrons in the atom (*not* the nucleus). The physical properties of the isotopes of an element (and compounds containing those isotopes) may differ because of the difference in mass of the isotopes.

10. Most elements are too reactive to be found in nature in other than the combined form. Aside from the noble metals gold, silver, and platinum, the only other elements commonly found in nature in the uncombined state are some of the gaseous elements (such as O_2, N_2, He, Ar, etc.), and the solid nonmetals carbon and sulfur.

12. Ionic compounds typically are hard, crystalline solids with high melting and boiling points. Ionic substances like sodium chloride, when dissolved in water or when melted, conduct electrical currents: chemists have taken this evidence to mean that ionic substances consist of positively and negatively charged particles (ions). Although an ionic substance is made up of positively and negatively charged particles, there is no net electrical charge on a sample of such a substance because the total number of positive charges is balanced by an equal number of negative charges. An ionic compound could not possibly exist of just cations or just anions: there must be a balance of charge or the compound would be very unstable (like charges repel each other).

14. When naming ionic compounds, we name the positive ion (cation) first. For simple binary Type I ionic compounds, the ending *-ide* is added to the root name of the element which is the negative ion (anion). For example, for the Type I ionic compound formed between potassium and sulfur, K_2S, the name would be potassium sulfide: potassium is the cation, sulfur is the anion (with the suffix *-ide* added). Type II compounds are named by either of two systems, the "*ous-ic*" system (which is falling out of use), and the "Roman numeral" system which is preferred by most chemists. Type II compounds involve elements which form more than one stable ion, and so it is necessary to specify *which* ion is present in a given compound. For example, iron forms two types of stable ion: Fe^{2+} and Fe^{3+}. Iron can react with oxygen to form either of two stable oxides, FeO or Fe_2O_3, depending on which cation is involved. Under the Roman numeral naming system, FeO would be named iron(II) oxide to show that it contains Fe^{2+} ions; Fe_2O_3 would be named iron(III) oxide to indicate that it contains Fe^{3+} ions. The Roman numeral used in a name corresponds to the charge of the specific ion present in the compound. Under the less-favored "*ous-ic*" system, for an element that forms two stable ions, the ending *-ous* is used to indicated the lower-charged ion, whereas the ending *-ic* is used to indicate the higher-charged ion. FeO and Fe_2O_3 would thus be named ferr*ous* oxide and ferr*ic* oxide, respectively. The "*ous-ic*" system has fallen out of favor since it does not indicate the actual charge on the ion, but only that it is the lower or higher charged of the two. This can lead to confusion: for example Fe^{2+} is called ferrous ion in this system, but Cu^{2+} is called cupric ion (since there is also a Cu^+ stable ion).

16. A polyatomic ion is an ion containing more than one atom. Some common polyatomic ions you should be familiar with are listed in Table 5.4. Parentheses are used in writing formulas containing polyatomic ions to indicate unambiguously how many of the polyatomic ion are present in the formula, to make certain that there is no mistake as to what is meant by the formula. For example, consider. For example, consider the substance calcium phosphate. The correct formula for this substance is $Ca_3(PO_4)_2$, which indicates that three calcium ions are combined for every two phosphate ions (check the total number of positive and negative charges to see why this is so). If we did not write the parenthesis around the formula for the phosphate ion, that is we had written Ca_3PO_{42}, people reading this formula might think that there were 42 oxygen atoms present!

18. Acids, in general, are substances which produce protons (H^+ ions) when dissolved in water. For acids which do not contain oxygen, the prefix *hydro-* and the suffix *-ic* are used with the root name of the element present in the acid (for example: HCl, hydrochloric acid; H_2S, hydrosulfuric acid; HF, hydrofluoric acid). The nomenclature of acids whose anions contain oxygen is more complicated. A series of prefixes and suffixes is used with the name of the central atom in the anion of the acid: these prefixes and suffixes indicate the relative (not actual) number of oxygen atoms present in the anion. Most of the elements that form oxyanions form two such anions: for example, sulfur forms sulfite ion (SO_3^{2-}) and sulfate ion (SO_4^{2-}), and nitrogen forms nitrite ion (NO_2^-) and nitrate ion (NO_3^-). For an element that forms two oxyanions, the acid containing the anions will have the ending *-ous* if the anion is the *-ite* anion and the ending *-ic* if the anion is the *-ate* anion. For example, HNO_2 is nitr*ous* acid and HNO_3 is nitr*ic* acid; H_2SO_3 is sulfur*ous* acid and H_2SO_4 is sulfur*ic* acid. The halogen elements (Group 7) each form four oxyanions, and consequently, four oxyacids. The prefix *hypo-* is used for the oxyacid that contains fewer oxygen atoms than the *-ite* anion, and the prefix *per-* is used for the oxyacid that contains more oxygen atoms than the *-ate* anion. For example,

acid	*name*	*anion*	*anion name*
HBrO	*hypo*brom*ous* acid	BrO^-	*hypo*brom*ite*
$HBrO_2$	brom*ous* acid	BrO_2^-	brom*ite*
$HBrO_3$	brom*ic* acid	BrO_3^-	brom*ate*
$HBrO_4$	*per*brom*ic* acid	BrO_4^-	*per*brom*ate*

20.

	Formula	Name	Atomic Number
a.	He	helium	2
b.	B	boron	5
c.	C	carbon	6
d.	F	fluorine	9

e.	S	sulfur	16
f.	Ba	barium	56
g.	Be	beryllium	4
h.	O	oxygen	8
i.	P	phosphorus	15
j.	Si	silicon	14

22.

		protons	neutrons	electrons
a.	$^{4}_{2}He$	2	2	2
b.	$^{37}_{17}Cl$	17	20	17
c.	$^{79}_{35}Br$	35	44	35
d.	$^{41}_{20}Ca$	20	21	20
e.	$^{40}_{20}Ca$	20	20	20
f.	$^{238}_{92}U$	92	146	92
g.	$^{235}_{92}U$	92	143	92
h.	$^{1}_{1}H$	1	0	1

24.
  a. K^+ (19 protons, 18 electrons)
  b. Ca^{2+} (20 protons, 18 electrons)
  c. N^{3-} (7 protons, 10 electrons)
  d. Br^- (35 protons, 36 electrons)
  e. Al^{3+} (13 protons, 10 electrons)
  f. Ag^+ (47 protons, 46 electrons)
  g. Cl^- (17 protons, 18 electrons)
  h. H^+ (1 proton, 0 electrons)
  i. H^- (1 proton, 2 electrons)

j. Na⁺ (11 protons, 10 electrons)
k. O²⁻ (8 protons, 10 electrons)
l. I⁻ (53 protons, 54 electrons)

26. a. FeCl₃, iron(III) chloride, ferric chloride
   b. Cu₂S, copper(I) sulfide, cuprous sulfide
   c. CoBr₂, cobalt(II) bromide, cobaltous bromide
   d. Fe₂O₃, iron(III) oxide, ferric oxide
   e. AuI₃, gold(III) iodide, auric iodide
   f. Cr₂S₃, chromium(III) sulfide, chromic sulfide
   g. MnO₂, manganese(IV) oxide, manganese dioxide (archaic)
   h. CuO, copper(II) oxide, cupric oxide
   i. NiS, nickel(II) sulfide, nickelous sulfide

28. a. NH₄⁺, ammonium ion
   b. SO₃²⁻, sulfite ion
   c. NO₃⁻, nitrate ion
   d. SO₄²⁻, sulfate ion
   e. NO₂⁻, nitrite ion
   f. CN⁻, cyanide ion
   g. OH⁻, hydroxide ion
   h. ClO₄⁻, perchlorate ion
   i. ClO⁻, hypochlorite ion
   j. PO₄³⁻, phosphate ion

30. a. B₂O₃, diboron trioxide
   b. NO₂, nitrogen dioxide
   c. PCl₅, phosphorus trichloride
   d. N₂O₄, dinitrogen tetroxide
   e. P₂O₅, diphosphorus pentoxide
   f. ICl, iodine monochloride
   g. SF₆, sulfur hexafluoride
   h. N₂O₃, dinitrogen trioxide

Chapter 6   Chemical Reactions: An Introduction

2. Hydrogen peroxide bubbles when it is applied to a wound. The bubbling represents the decomposition of the hydrogen peroxide into oxygen gas (the bubbles) and water.

4. The oils and greases deposited on the walls of the oven are converted by the sodium hydroxide into soap, which is water soluble and can be washed away. The formation of soap represents the chemical breakdown of greases and oils.

6. A gas (carbon dioxide) is evolved during fermentation and the bubbles may be noticed. The odor of the product of the fermentation (alcohol or vinegar) will be noticeable as the concentration of product builds up.

8. atoms

10. the same

12. water

14. $Zn(s) + S(s) \rightarrow ZnS(s)$

16. $(NH_4)_2CO_3(s) \rightarrow NH_3(g) + CO_2(g) + H_2O(g)$

18. $Cl_2(g) + KI(aq) \rightarrow KCl(aq) + I_2(s)$

20. $Ca(s) + H_2O(l) \rightarrow Ca(OH)_2(s) + H_2(g)$

22. $CaC_2(s) + H_2O(l) \rightarrow C_2H_2(g) + Ca(OH)_2(s)$

24. $H_2S(g) + O_2(g) \rightarrow SO_2(g) + H_2O(g)$

26. $NaCl(aq) + H_2O(l) \rightarrow NaOH(aq) + Cl_2(g) + H_2(g)$

28. $O_2(g) \rightarrow O_3(g)$

30. $SiO_2(s) + C(s) \rightarrow Si(s) + CO(g)$

32. $Xe(g) + F_2(g) \rightarrow XeF_4(s)$

34. $C_{12}H_{22}O_{11}(s) + O_2(g) \rightarrow CO_2(g) + H_2O(g)$

36. whole numbers

38. a. $Br_2 + KI \rightarrow KBr + I_2$

   Balance bromine: $Br_2 + KI \rightarrow 2KBr + I_2$

   Balance iodine: $Br_2 + 2KI \rightarrow 2KBr + I_2$

   Balanced equation: $Br_2(l) + 2KI(aq) \rightarrow 2KBr(aq) + I_2(s)$

   b. $Co + O_2 \rightarrow Co_2O_3$

   Equations of this sort can be difficult to balance. In Co_2O_3, we have three oxygen atoms, whereas in O_2 we have two oxygen atoms. No simple multiple of either substance *alone* can lead to the same number of oxygen atoms on both sides of the equation. We need to find the smallest whole number that *both* three and two divide into. If we take *two* Co_2O_3, this will contain six oxygen atoms. If we take *three* O_2, this will also contain six oxygen atoms. Six is the smallest number that both two and three divide into (least common multiple).

Balance oxygen: Co + 3O$_2$ → 2Co$_2$O$_3$

Balance cobalt: 4Co + 3O$_2$ → 2Co$_2$O$_3$

Balanced equation: 4Co(s) + 3O$_2$(g) → 2Co$_2$O$_3$(s)

c. P$_4$ + O$_2$ → P$_2$O$_5$

Balance phosphorus: P$_4$ + O$_2$ → 2P$_2$O$_5$

Balance oxygen: P$_4$ + 5O$_2$ → 2P$_2$O$_5$

Balanced equation: P$_4$(s) + 5O$_2$(g) → 2P$_2$O$_5$(s)

d. Al + HNO$_3$ → Al(NO$_3$)$_3$ + H$_2$

Hint: since nitrate ions are present on each side of the equation [in HNO$_3$ and in Al(NO$_3$)$_3$], and do not break down during the reaction, balance nitrate ions as a single unit (not as individual nitrogen and oxygen atoms).

Balance nitrate ions: Al + 3HNO$_3$ → Al(NO$_3$)$_3$ + H$_2$

Balance hydrogen: Al + 6HNO$_3$ → Al(NO$_3$)$_3$ + 3H$_2$

Rebalance nitrate ions: Al + 6HNO$_3$ → 2Al(NO$_3$)$_3$ + 3H$_2$

Balance aluminum: 2Al + 6HNO$_3$ → 2Al(NO$_3$)$_3$ + 3H$_2$

Balanced equation: 2Al(s) + 6HNO$_3$(aq) → 2Al(NO$_3$)$_3$(aq) + 3H$_2$(g)

e. PBr$_3$ + H$_2$O → H$_3$PO$_3$ + HBr

Balance bromine: PBr$_3$ + H$_2$O → H$_3$PO$_3$ + 3HBr

Balance hydrogen: PBr$_3$ + 3H$_2$O → H$_3$PO$_3$ + 3HBr

Balanced equation: PBr$_3$(l) + 3H$_2$O(l) → H$_3$PO$_3$(aq) + 3HBr(aq)

f. NO + O$_2$ → NO$_2$

This equation looks so simple, but can be hard to balance by inspection. The left side of the equation has three oxygen atoms, while the right side has only two oxygen atoms. If we had *individual* oxygen atoms available, balancing would be simple, but oxygen occurs naturally as O$_2$ molecules, which are all we can use when balancing oxygen atoms (you cannot change the formulas of any of the reactants or products when balancing an equation). If we put a coefficient of 2 in front of each of the nitrogen compounds, however, we go from having a difference of *one* oxygen atom between the sides of the equation to having a difference of *two* oxygen atoms, which makes for one O$_2$ molecule: 2NO(g) + O$_2$(g) → 2NO$_2$(g)

Balanced equation: 2NO(g) + O$_2$(g) → 2NO$_2$(g)

g. 2C$_2$H$_6$ + 7O$_2$ → 4CO$_2$ + 6H$_2$O

Hint: equations involving the reaction with O$_2$ of carbon/hydrogen compounds in which there is an *even* number of carbon atoms very often require a coefficient of 2 for the carbon/hydrogen compound in order to balance the number of oxygen atoms used.

Balance hydrogen: $C_2H_6 + O_2 \rightarrow CO_2 + 3H_2O$

Balance carbon: $C_2H_6 + O_2 \rightarrow 2CO_2 + 3H_2O$

Notice that seven oxygen atoms are needed to balance the number of oxygen atoms on the right side of the equation, but that oxygen occurs on the left side of the equation as O_2 molecules, which cannot be changed. If a coefficient of 2, however, is placed in front of C_2H_6, and carbon and hydrogen are rebalanced, a whole number of O_2 molecules will be needed.

Add coefficient of 2 for C_2H_6: $2C_2H_6 + O_2 \rightarrow 2CO_2 + 3H_2O$

Rebalance carbon: $2C_2H_6 + O_2 \rightarrow 4CO_2 + 3H_2O$

Rebalance hydrogen $2C_2H_6 + O_2 \rightarrow 4CO_2 + 6H_2O$

Balance oxygen: $2C_2H_6 + 7O_2 \rightarrow 4CO_2 + 6H_2O$

Balanced equation: $2C_2H_6(g) + 7O_2(g) \rightarrow 4CO_2(g) + 6H_2O(g)$

h. $CuO + H_2SO_4 \rightarrow CuSO_4 + H_2O$ This equation is already balanced!

40. a. $CaF_2 + H_2SO_4 \rightarrow CaSO_4 + HF$

Balance fluorine: $CaF_2 + H_2SO_4 \rightarrow CaSO_4 + 2HF$

Balanced equation: $CaF_2(aq) + H_2SO_4(aq) \rightarrow CaSO_4(s) + 2HF(aq)$

b. $KBr + H_3PO_4 \rightarrow K_3PO_4 + HBr$

Balance potassium: $3KBr + H_3PO_4 \rightarrow K_3PO_4 + HBr$

Balance bromine: $3KBr + H_3PO_4 \rightarrow K_3PO_4 + 3HBr$

Balanced equation: $3KBr(aq) + H_3PO_4(aq) \rightarrow K_3PO_4(aq) + 3HBr(g)$

c. $TiCl_4 + Na \rightarrow NaCl + Ti$

Balance chlorine: $TiCl_4 + Na \rightarrow 4NaCl + Ti$

Balance sodium: $TiCl_4 + 4Na \rightarrow 4NaCl + Ti$

Balanced equation: $TiCl_4(l) + 4Na(s) \rightarrow 4NaCl(s) + Ti(s)$

d. $K_2CO_3 \rightarrow K_2O + CO_2$ This equation is already balanced!

e. $KO_2 + H_2O \rightarrow KOH + O_2$

Balance hydrogen: $KO_2 + H_2O \rightarrow 2KOH + O_2$

Balance potassium: $2KO_2 + H_2O \rightarrow 2KOH + O_2$

At this point, we have balanced potassium and hydrogen atoms, but now it becomes very difficult to balance oxygen since it occurs in each of the reactants and products. There is no simple, systematic way to balance oxygen at this point (you will learn a special method in a later chapter for oxidation-reduction reactions such as this). We need one more oxygen atom on the right side of the equation to balance it: if we could just have an extra half of an O_2 molecule (that is, $1.5O_2$) the equation would be balanced. That is

Balance using non-integer: $2KO_2 + H_2O \rightarrow 2KOH + 1.5O_2$

Although this equation is balanced, we can't really have 1.5 molecules. If we multiply everything in this equation by 2, however, we will get whole number coefficients.

Balanced equation: $4KO_2(s) + 2H_2O(l) \rightarrow 4KOH(aq) + 3O_2(g)$

f. $Na_2O_2 + H_2O + CO_2 \rightarrow NaHCO_3 + O_2$

Balance sodium: $Na_2O_2 + H_2O + CO_2 \rightarrow 2NaHCO_3 + O_2$

Balance carbon: $Na_2O_2 + H_2O + 2CO_2 \rightarrow 2NaHCO_3 + O_2$

Again we are left with the difficulty of balancing oxygen when it occurs in all the reactants and products. Right now, there are 7 oxygen atoms on the left side of the equation, but there are 8 oxygen atoms on the right side. If we could have just half an O_2 molecule (instead of one entire molecule) things would work out.

Balance using non-integer: $Na_2O_2 + H_2O + 2CO_2 \rightarrow 2NaHCO_3 + 0.5O_2$

Although this equation is balanced, we can't really have half a molecule. If we multiply everything in the equation by 2, however, we'll get whole numbers.

Balanced equation: $2Na_2O_2(s) + 2H_2O(g) + 4CO_2(g) \rightarrow 4NaHCO_3(s) + O_2(g)$

g. $KNO_2 + C \rightarrow K_2CO_3 + CO + N_2$

Balance nitrogen: $2KNO_2 + C \rightarrow K_2CO_3 + CO + N_2$

Balance carbon: $2KNO_2 + 2C \rightarrow K_2CO_3 + CO + N_2$

Balanced equation: $2KNO_2(s) + 2C(s) \rightarrow K_2CO_3(s) + CO(g) + N_2(g)$

h. $BaO + Al \rightarrow Ba + Al_2O_3$

Balance aluminum: $BaO + 2Al \rightarrow Ba + Al_2O_3$

Balance oxygen: $3BaO + 2Al \rightarrow Ba + Al_2O_3$

Balance barium: $3BaO + 2Al \rightarrow 3Ba + Al_2O_3$

Balanced equation: $3BaO(s) + 2Al(s) \rightarrow 3Ba(s) + Al_2O_3(s)$

42. a. $SiI_4(s) + 2Mg(s) \rightarrow Si(s) + 2MgI_2(s)$
    b. $MnO_2(s) + 2Mg(s) \rightarrow Mn(s) + 2MgO(s)$
    c. $8Ba(s) + S_8(s) \rightarrow 8BaS(s)$
    d. $4NH_3(g) + 3Cl_2(g) \rightarrow 3NH_4Cl(s) + NCl_3(g)$
    e. $8Cu_2S(s) + S_8(s) \rightarrow 16CuS(s)$
    f. $2Al(s) + 3H_2SO_4(aq) \rightarrow Al_2(SO_4)_3(aq) + 3H_2(g)$

Chapter 6 Chemical Reactions: An Introduction

    g. $2NaCl(s) + H_2SO_4(l) \rightarrow 2HCl(g) + Na_2SO_4(s)$

    h. $2CO(g) + O_2(g) \rightarrow 2CO_2(g)$

44. a. $Ba(NO_3)_2(aq) + Na_2CrO_4(aq) \rightarrow BaCrO_4(s) + 2NaNO_3(aq)$

    b. $PbCl_2(aq) + K_2SO_4(aq) \rightarrow PbSO_4(s) + 2KCl(aq)$

    c. $C_2H_5OH(l) + 3O_2(g) \rightarrow 2CO_2(g) + 3H_2O(l)$

    d. $CaC_2(s) + 2H_2O(l) \rightarrow Ca(OH)_2(s) + H_2C_2(g)$

    e. $Sr(s) + 2HNO_3(aq) \rightarrow Sr(NO_3)_2(aq) + H_2(g)$

    f. $BaO_2(s) + H_2SO_4(aq) \rightarrow BaSO_4(s) + H_2O_2(aq)$

    g. $2AsI_3(s) \rightarrow 2As(s) + 3I_2(s)$

    h. $2CuSO_4(aq) + 4KI(s) \rightarrow 2CuI(s) + I_2(s) + 2K_2SO_4(aq)$

46. The beauty of the periodic chart is, if you know how *one* element of a group reacts, then you can be almost certain that all the other members of the group will react in a similar fashion. This permits you to write a "group reaction" that summarizes the properties of the entire group. For example, if M represents a typical alkali metal (Group 1), then the unbalanced reaction of M with water can be written as

$$2M + 2H_2O \rightarrow 2MOH + H_2$$

This reaction applies for all the metals of Group 1:

$2Li(s) + 2H_2O(l) \rightarrow 2LiOH(aq) + H_2(g)$

$2Na(s) + 2H_2O(l) \rightarrow 2NaOH(aq) + H_2(g)$

$2K(s) + 2H_2O(l) \rightarrow 2KOH(aq) + H_2(g)$

$2Rb(s) + 2H_2O(l) \rightarrow 2RbOH(aq) + H_2(g)$

$2Cs(s) + 2H_2O(l) \rightarrow 2CsOH(aq) + H_2(g)$

$2Fr(s) + 2H_2O(l) \rightarrow 2FrOH(aq) + H_2(g)$

Notice the correspondence between these specific reactions and the general group reaction for M.

48. $C_{12}H_{22}O_{11}(aq) + H_2O(l) \rightarrow 4C_2H_6O(aq) + 4CO_2(g)$

50. $2Al_2O_3(s) + 3C(s) \rightarrow 4Al(s) + 3CO_2(g)$

52. Let M represent a Group 2 metal, and X_2 represent the halogens; then

$$M + X_2 \rightarrow MX_2$$

represents the general group reaction. The specific reactions are

$Be(s) + F_2(g) \rightarrow BeF_2(s)$

$Be(s) + Cl_2(g) \rightarrow BeCl_2(s)$

$Mg(s) + F_2(g) \rightarrow MgF_2(s)$

$Mg(s) + Cl_2(g) \rightarrow MgCl_2(s)$

$Ca(s) + F_2(g) \rightarrow CaF_2(s)$

$Ca(s) + Cl_2(g) \rightarrow CaCl_2(s)$

$Sr(s) + F_2(g) \rightarrow SrF_2(s)$

$Sr(s) + Cl_2(g) \rightarrow SrCl_2(s)$

$Ba(s) + F_2(g) \rightarrow BaF_2(s)$

$Ba(s) + Cl_2(g) \rightarrow BaCl_2(s)$

$Ra(s) + F_2(g) \rightarrow RaF_2(s)$

$Ra(s) + Cl_2(g) \rightarrow RaCl_2(s)$

54. $(NH_4)_2Cr_2O_7(s) \rightarrow Cr_2O_3(s) + N_2(g) + 4H_2O(g)$

56. $2KClO_3(s) \rightarrow 2KCl(s) + 3O_2(g)$

58. $NH_3(g) + HCl(g) \rightarrow NH_4Cl(s)$

60. Most oils *decompose* at high temperatures, producing a foul-smelling chemical called acrolein that we associate with the smell and taste of burned food. Oil that has been heated to too high a temperature for too long a period "turns rancid" and is unfit for further use. Also, at high enough temperatures, the oil might ignite.

62. $Fe(s) + S(s) \rightarrow FeS(s)$

64. $C_3H_8(g) + O_2(g) \rightarrow CO_2(g) + H_2O(g)$

66. $KI(aq) + H_2O(l) \rightarrow KOH(aq) + H_2(g) + I_2(s)$

68. $Mg(s) + H_2O(g) \rightarrow Mg(OH)_2(s) + H_2(g)$

70. $CuO(s) + H_2SO_4(aq) \rightarrow CuSO_4(aq) + H_2O(l)$

72. $Na_2SO_3(aq) + S(s) \rightarrow Na_2S_2O_3(aq)$

74. a. $Ba(NO_3)_2 + KF \rightarrow BaF_2 + KNO_3$

   Balance nitrate ion: $Ba(NO_3)_2 + KF \rightarrow BaF_2 + 2KNO_3$

   Balance potassium: $Ba(NO_3)_2 + 2KF \rightarrow BaF_2 + 2KNO_3$

   Fluorine and barium are now balanced also.

   Balanced equation: $Ba(NO_3)_2(aq) + 2KF(aq) \rightarrow BaF_2(s) + 2KNO_3(aq)$

   b. $Zn + HCl \rightarrow ZnCl_2 + H_2$

   Balance hydrogen: $Zn + 2HCl \rightarrow ZnCl_2 + H_2$

   Chlorine is also now balanced.

   Balanced equation: $Zn(s) + 2HCl(aq) \rightarrow ZnCl_2(aq) + H_2(g)$

   c. $Fe + S \rightarrow Fe_2S_3$

   Balance Fe: $2Fe + S \rightarrow Fe_2S_3$

54 Chapter 6 Chemical Reactions: An Introduction

      Balance S: $2Fe + 3S \rightarrow Fe_2S_3$

      Balanced equation: $2Fe(s) + 3S(s) \rightarrow Fe_2S_3(s)$

d. $C_6H_{12}O_6 + O_2 \rightarrow CO_2 + H_2O$

      Balance carbon: $C_6H_{12}O_6 + O_2 \rightarrow 6CO_2 + H_2O$

      Balance hydrogen: $C_6H_{12}O_6 + O_2 \rightarrow 6CO_2 + 6H_2O$

      Balance oxygen: $C_6H_{12}O_6 + 6O_2 \rightarrow 6CO_2 + 6H_2O$

      Balanced equation: $C_6H_{12}O_6(s) + 6O_2(g) \rightarrow 6CO_2(g) + 6H_2O(g)$

e. $H_2O + Cl_2 \rightarrow HCl + HOCl$ This equation is already balanced!

f. $ZnS + O_2 \rightarrow ZnO + SO_2$

Notice that the right side of the equation has *three* oxygen atoms, whereas the left side has only *two* oxygen atoms. The only source of oxygen atoms on the left side of the equation is as O_2 molecules. We will have to end up with an *even number* of oxygen atoms on the right side of the equation, since we cannot change the formula of O_2. So try adding a coefficient of 2 for ZnO to make an even number of oxygen atoms on the right side

      $ZnS + O_2 \rightarrow 2ZnO + SO_2$

      Balance zinc: $2ZnS + O_2 \rightarrow 2ZnO + SO_2$

      Balance sulfur: $2ZnS + O_2 \rightarrow 2ZnO + 2SO_2$

      Balance oxygen: $2ZnS + 3O_2 \rightarrow 2ZnO + 2SO_2$

      Balanced equation: $2ZnS(s) + 3O_2(g) \rightarrow 2ZnO(s) + 2SO_2(g)$

g. $PbSO_4 + NaCl \rightarrow Na_2SO_4 + Na_2PbCl_4$

      Balance chlorine: $PbSO_4 + 4NaCl \rightarrow Na_2SO_4 + Na_2PbCl_4$

      All other atoms are now balanced.

      Balanced equation: $PbSO_4(s) + 4NaCl(aq) \rightarrow Na_2SO_4(aq) + Na_2PbCl_4(aq)$

h. $Fe_2O_3 + C \rightarrow Fe_3O_4 + CO$

Notice that we have two iron atoms on the *left* side of the equation and three iron atoms on the *right* side. There is no simple multiple of two or three that will give the right number of iron atoms for the opposite side of the equation. The smallest whole number that is divisible by both 2 and 3 is *six*, so the coefficients should be adjusted so that there are six iron atoms on each side of the equation

      Balance iron: $3Fe_2O_3 + C \rightarrow 2Fe_3O_4 + CO$

      All other atoms are now balanced.

      Balanced equation: $3Fe_2O_3(s) + C(s) \rightarrow 2Fe_3O_4(s) + CO(g)$

76. a. $Pb(NO_3)_2(aq) + K_2CrO_4(aq) \rightarrow PbCrO_4(s) + 2KNO_3(aq)$
    b. $BaCl_2(aq) + Na_2SO_4(aq) \rightarrow BaSO_4(s) + 2NaCl(aq)$
    c. $2CH_3OH(l) + 3O_2(g) \rightarrow 2CO_2(g) + 4H_2O(g)$
    d. $Na_2CO_3(aq) + S(s) + SO_2(g) \rightarrow CO_2(g) + Na_2S_2O_3(aq)$
    e. $Cu(s) + 2H_2SO_4(aq) \rightarrow CuSO_4(aq) + SO_2(g) + 2H_2O(l)$
    f. $MnO_2(s) + 4HCl(aq) \rightarrow MnCl_2(aq) + Cl_2(g) + 2H_2O(l)$
    g. $As_2O_3(s) + 6KI(aq) + 6HCl(aq) \rightarrow 2AsI_3(s) + 6KCl(aq) + 3H_2O(l)$
    h. $2Na_2S_2O_3(aq) + I_2(aq) \rightarrow Na_2S_4O_6(aq) + 2NaI(aq)$

Chapter 7   Reactions in Aqueous Solutions

2. Driving forces are types of *changes* in a system which pull a reaction in the *direction of product formation*; driving forces discussed in Chapter Seven include: formation of a *solid*, formation of *water*, formation of a *gas*, transfer of electrons.

4. The net charge of a precipitate must be *zero*. The total number of positive charges equals the total number of negative charges.

6. ions

8. Consider the two new possible combinations of ions (when two ionic compounds are mixed the ions may "switch partners"); if one of these new possible combinations is insoluble in water it will form a precipitate having zero net charge (indicating in what proportions the ions must be combined). There is no foolproof method for predicting by inspection what precipitate may form: we rely on intuition, experience from similar reactions, and "rules of thumb" based on the results of experiments (such as the general solubility rules from Table 7.1).

10. For most practical purposes, "insoluble" and "slightly" soluble mean the same thing. The difference between "insoluble" and "slightly soluble" could be crucial if, for example, a substance were highly toxic and were found in a water supply.

12.
    a. soluble (Rule 1: most nitrate salts are soluble)
    b. soluble (Rule 2: most potassium salts are soluble)
    c. soluble (Rule 2: most sodium salts are soluble)
    d. insoluble (Rule 5: most hydroxide compounds are insoluble)
    e. insoluble (Rule 3: exception for chloride salts)
    f. soluble (Rule 2: most ammonium salts are soluble)
    g. insoluble (Rule 6: most sulfide salts are insoluble)
    h. insoluble (Rule 4: exception for sulfate salts)

14.
    a. Rule 6: most sulfide salts are only slightly soluble
    b. Rule 5: most hydroxide compounds are only slightly soluble
    c. Rule 4: exception
    d. Rule 6: most carbonate salts are only slightly soluble

16.
    a. mercury(II) sulfide, HgS. Rule 6: most sulfide salts are only slightly soluble.
    b. calcium sulfate, $CaSO_4$. Rule 4: $CaSO_4$ is a listed exception.
    c. silver chloride, AgCl. Rule 3: AgCl is a listed exception.

d. barium carbonate, $BaCO_3$. Rule 6: most carbonate compounds are only slightly soluble.

e. nickel(II) phosphate, $Ni_3(PO_4)_2$. Rule 6: most phosphate salts are only slightly soluble.

f. copper(II) carbonate, $CuCO_3$. Rule 6: most carbonate salts are only slightly soluble.

18. The precipitates are marked in boldface type.

   a. No precipitate: $Ba(NO_3)_2$ and HCl are each soluble.

   b. Rule 6: most sulfide salts are insoluble.

   $(NH_4)_2S(aq) + CoCl_2(aq) \rightarrow \textbf{CoS}(s) + 2NH_4Cl(aq)$

   c. Rule 4: lead sulfate is a listed exception.

   $H_2SO_4(aq) + Pb(NO_3)_2(aq) \rightarrow \textbf{PbSO}_4(s) + 2HNO_3(aq)$

   d. Rule 6: most carbonate salts are insoluble.

   $CaCl_2(aq) + H_2CO_3(aq) \rightarrow \textbf{CaCO}_3(s) + 2HCl(aq)$

   e. Rule 6: most phosphate salts are insoluble

   $Na_3PO_4(aq) + CrCl_3(aq) \rightarrow 3NaCl(aq) + \textbf{CrPO}_4(s)$

20. Hint: when balancing equations involving polyatomic ions, especially in precipitation reactions, balance the polyatomic ions as a *unit*, not in terms of the atoms the polyatomic ions contain (e.g., treat nitrate ion, NO_3^- as a single entity, not as one nitrogen and three oxygen atoms). When finished balancing, however, do be sure to count the individual number of atoms of each type on each side of the equation.

   a. $AgNO_3(aq) + H_2SO_4(aq) \rightarrow Ag_2SO_4(s) + HNO_3(aq)$

   Balance silver: $2AgNO_3(aq) + H_2SO_4(aq) \rightarrow Ag_2SO_4(s) + HNO_3(aq)$

   Balance nitrate: $2AgNO_3(aq) + H_2SO_4(aq) \rightarrow Ag_2SO_4(s) + 2HNO_3(aq)$

   Balanced equation: $2AgNO_3(aq) + H_2SO_4(aq) \rightarrow Ag_2SO_4(s) + 2HNO_3(aq)$

   b. $Ca(NO_3)_2(aq) + H_2SO_4(aq) \rightarrow CaSO_4(s) + HNO_3(aq)$

   Balance nitrate: $Ca(NO_3)_2(aq) + H_2SO_4(aq) \rightarrow CaSO_4(s) + 2HNO_3(aq)$

   Balanced equation: $Ca(NO_3)_2(aq) + H_2SO_4(aq) \rightarrow CaSO_4(s) + 2HNO_3(aq)$

   c. $Pb(NO_3)_2(aq) + H_2SO_4(aq) \rightarrow PbSO_4(s) + HNO_3(aq)$

   Balance nitrate: $Pb(NO_3)_2(aq) + H_2SO_4(aq) \rightarrow PbSO_4(s) + 2HNO_3(aq)$

   Balanced equation: $Pb(NO_3)_2(aq) + H_2SO_4(aq) \rightarrow PbSO_4(s) + 2HNO_3(aq)$

22. The products are determined by having the ions "switch partners." For example, for a general reaction AB + CD →, the possible products are AD and CB if the ions switch partners. If either AD or CB is insoluble,

then a precipitation reaction has occurred. In the following reaction, the formula of the precipitate is given in boldface type.

a. $NiCl_2(aq) + H_2S(aq) \rightarrow $ **$NiS(s)$** $+ 2HCl(aq)$

   Rule 6: most sulfide salts are only slightly soluble.

b. $CuSO_4(aq) + 2NaOH(aq) \rightarrow Na_2SO_4(aq) + $ **$Cu(OH)_2(s)$**

   Rule 5: most hydroxide compounds are only slightly soluble.

c. $Ba(NO_3)_2(aq) + Na_2CO_3(aq) \rightarrow 2NaNO_3(aq) + $ **$BaCO_3(s)$**

   Rule 6: most carbonate salts are only slightly soluble.

24. spectator

26. The net ionic equation for a reaction indicates *only those ions that go to form the precipitate*, and does not show the spectator ions present in the solutes mixed. The identity of the precipitate is determined from the Solubility Rules (Table 7.1).

a. $Fe^{3+}(aq) + 3OH^-(aq) \rightarrow Fe(OH)_3(s)$

   Rule 5: most hydroxide compounds are only slightly soluble.

b. $Ni^{2+}(aq) + S^{2-}(aq) \rightarrow NiS(s)$

   Rule 6: most sulfide salts are only slightly soluble.

c. $Ag^+(aq) + Cl^-(aq) \rightarrow AgCl(s)$

   Rule 3: AgCl is a listed exception.

d. $Ba^{2+}(aq) + SO_4^{2-}(aq) \rightarrow BaSO_4(s)$

   Rule 4: $BaSO_4$ is a listed exception.

e. $Hg_2^{2+}(aq) + 2Br^-(aq) \rightarrow Hg_2Br_2(s)$

   Rule 3: Br and Cl are both in Group 7; Hg_2Br_2 is insoluble also.

f. $Ba^{2+}(aq) + SO_4^{2-}(aq) \rightarrow BaSO_4(s)$

   Rule 4: $BaSO_4$ is a listed exception.

28. $Ag^+(aq) + Cl^-(aq) \rightarrow AgCl(s)$

   $Pb^{2+}(aq) + 2Cl^-(aq) \rightarrow PbCl_2(s)$

   $Hg_2^{2+}(aq) + 2Cl^-(aq) \rightarrow Hg_2Cl_2(s)$

30. $Co^{2+}(aq) + S^{2-}(aq) \rightarrow CoS(s)$

   $2Co^{3+}(aq) + 3S^{2-}(aq) \rightarrow Co_2S_3(s)$

   $Fe^{2+}(aq) + S^{2-}(aq) \rightarrow FeS(s)$

   $2Fe^{3+}(aq) + 3S^{2-}(aq) \rightarrow Fe_2S_3(s)$

Chapter 7   Reactions in Aqueous Solutions   59

32. Acids were noted very early in history for their sour taste (for example, the acetic acid in vinegar and the citric acid in lemon juice). Bases were noted because of their bitter taste and slippery feel on the skin.

34. HCl

36. salt

38. $RbOH(s) \rightarrow Rb^+(aq) + OH^-(aq)$

$CsOH(s) \rightarrow Cs^+(aq) + OH^-(aq)$

40. In general, the salt formed in an aqueous acid-base reaction consists of the *positive ion of the base* involved in the reaction, combined with the *negative ion of the acid*. The hydrogen ion of the strong acid combines with the hydroxide ion of the strong base to produce water, which is the other product of the acid-base reactions.

   a. $NaOH(aq) + HNO_3(aq) \rightarrow NaNO_3(aq) + H_2O(l)$
   b. $2KOH(aq) + H_2SO_4(aq) \rightarrow K_2SO_4(aq) + 2H_2O(l)$
   c. $RbOH(aq) + HCl(aq) \rightarrow RbCl(aq) + H_2O(l)$
   d. $NaOH(aq) + HClO_4(aq) \rightarrow NaClO_4(aq) + H_2O(l)$

42. A *molecular equation* uses the normal, uncharged formulas for the compounds involved. The *complete ionic equation* shows the compounds involved broken up into their respective ions (*all* ions present are shown). The *net ionic equation* shows only those ions which combine to form a precipitate, a gas, or a nonionic product such as water. The net ionic equation shows most clearly the species that are combining with each other.

44. $Pb^{2+}(aq) + 2Cl^-(aq) \rightarrow PbCl_2(s)$

$Pb^{2+}(aq) + CrO_4^{2-}(aq) \rightarrow PbCrO_4(s)$

46. The formulas of the precipitates are predicted using the Solubility Rules of Table 7.1. You should be able by now to predict the formation of a precipitate without having to write out full molecular or ionic equations.

   a. Rule 5: $Co^{3+}(aq) + 3OH^-(aq) \rightarrow Co(OH)_3(s)$
   b. Rule 6: $2Ag^+(aq) + CO_3^{2-}(aq) \rightarrow Ag_2CO_3(s)$
   c. no reaction (all combinations of ions are soluble)
   d. Rule 4: $Ba^{2+}(aq) + SO_4^{2-}(aq) \rightarrow BaSO_4(s)$
   e. no reaction (all combinations of ions are soluble)
   f. Rule 6: $3Ca^{2+}(aq) + 2PO_4^{3-}(aq) \rightarrow Ca_3(PO_4)_2(s)$
   g. Rule 5: $Al^{3+}(aq) + 3OH^-(aq) \rightarrow Al(OH)_3(s)$

60   Chapter 7   Reactions in Aqueous Solutions

48. $Ca(OH)_2(s) \rightarrow Ca^{2+}(aq) + 2OH^-(aq)$

$Mg(OH)_2(s) \rightarrow Mg^{2+}(aq) + 2OH^-(aq)$

$Sr(OH)_2(s) \rightarrow Sr^{2+}(aq) + 2OH^-(aq)$

$Ba(OH)_2(s) \rightarrow Ba^{2+}(aq) + 2OH^-(aq)$

50. When *any* strong acid is reacted with *any* strong base, the net ionic reaction is the *same*:

$H^+(aq) + OH^-(aq) \rightarrow H_2O(l)$

Since the net ionic reaction is the same for any strong acid, the amount of heat liberated is also the same.

52. a. soluble (Rule 2: most potassium salts are soluble)

b. soluble (Rule 2: most ammonium salts are soluble)

c. insoluble (Rule 6: most carbonate salts are only slightly soluble)

d. insoluble (Rule 6: most phosphate salts are only slightly soluble)

e. soluble (Rule 2: most sodium salts are soluble)

f. insoluble (Rule 6: most carbonate salts are only slightly soluble)

g. soluble (Rule 3: most chloride salts are soluble)

54. a. iron(III) hydroxide, $Fe(OH)_3$. Rule 5: most hydroxide salts are only slightly soluble

b. nickel(II) sulfide, NiS. Rule 6: most sulfide salts are only slightly soluble

c. silver chloride, AgCl. Rule 3: Although most chloride salts are soluble, AgCl is a listed exception

d. barium carbonate, $BaCO_3$. Rule 6: most carbonate salts are only slightly soluble

e. mercury(I) chloride or mercurous chloride, Hg_2Cl_2. Rule 3: Although most chloride salts are soluble, Hg_2Cl_2 is a listed exception

f. barium sulfate, $BaSO_4$. Rule 4: Although most sulfate salts are soluble, $BaSO_4$ is a listed exception

56. Hint: when balancing equations involving polyatomic ions, especially in precipitation reactions, balance the polyatomic ions as a *unit*, not in terms of the atoms the polyatomic ions contain (e.g., treat nitrate ion, NO_3^- as a single entity, not as one nitrogen and three oxygen atoms). When finished balancing, however, do be sure to count the individual number of atoms of each type on each side of the equation.

Chapter 7   Reactions in Aqueous Solutions   61

a. $Co(NO_3)_2(aq) + (NH_4)_2S(aq) \rightarrow CoS(s) + NH_4NO_3(aq)$

Balance NO_3^- ions: $Co(NO_3)_2(aq) + (NH_4)_2S(aq) \rightarrow CoS(s) + 2NH_4NO_3(aq)$

Equation is now balanced.

b. $CaCl_2(aq) + H_2SO_4(aq) \rightarrow CaSO_4(s) + HCl(aq)$

Balance chlorine: $CaCl_2(aq) + H_2SO_4(aq) \rightarrow CaSO_4(s) + 2HCl(aq)$

Equation is now balanced.

c. $FeCl_3(aq) + Na_3PO_4(aq) \rightarrow FePO_4(s) + NaCl(aq)$

Balance chlorine: $FeCl_3(aq) + Na_3PO_4(aq) \rightarrow FePO_4(s) + 3NaCl(aq)$

Equation is now balanced.

58.
a. Rule 3: $Ag^+(aq) + Cl^-(aq) \rightarrow AgCl(s)$
b. Rule 6: $3Ca^{2+}(aq) + 2PO_4^{3-}(aq) \rightarrow Ca_3(PO_4)_2(s)$
c. Rule 3: $Pb^{2+}(aq) + 2Cl^-(aq) \rightarrow PbCl_2(s)$
d. Rule 6: $Fe^{3+}(aq) + 3OH^-(aq) \rightarrow Fe(OH)_3(s)$

60. $Fe^{2+}(aq) + S^{2-}(aq) \rightarrow FeS(s)$

$2Cr^{3+}(aq) + 3S^{2-}(aq) \rightarrow Cr_2S_3(s)$

$Ni^{2+}(aq) + S^{2-}(aq) \rightarrow NiS(s)$

62. hydroxide

64. The formulas of the salts are marked in boldface type. Remember that in an acid/base reaction in aqueous solution, *water* is always one of the products: keeping this in mind makes predicting the formula of the *salt* produced easy to do.

a. $HClO_4(aq) + RbOH(aq) \rightarrow H_2O(l) + \mathbf{RbClO_4}(aq)$
b. $HNO_3(aq) + KOH(aq) \rightarrow H_2O(l) + \mathbf{KNO_3}(aq)$
c. $H_2SO_4(aq) + 2NaOH(aq) \rightarrow 2H_2O(l) + \mathbf{Na_2SO_4}(aq)$
d. $HBr(aq) + CsOH(aq) \rightarrow H_2O(l) + \mathbf{CsBr}(aq)$

66. These anions tend to form insoluble precipitates with *many* metal ions. The following are illustrative for cobalt(II) chloride, tin(II) chloride, and copper(II) nitrate reacting with the sodium salts of the given anions.

a. $CoCl_2(aq) + Na_2S(aq) \rightarrow CoS(s) + 2NaCl(aq)$

$SnCl_2(aq) + Na_2S(aq) \rightarrow SnS(s) + 2NaCl(aq)$

$Cu(NO_3)_2(aq) + Na_2S(aq) \rightarrow CuS(s) + 2NaNO_3(aq)$

b. $CoCl_2(aq) + Na_2CO_3(aq) \rightarrow CoCO_3(s) + 2NaCl(aq)$

$SnCl_2(aq) + Na_2CO_3(aq) \rightarrow SnCO_3(s) + 2NaCl(aq)$

$Cu(NO_3)_2(aq) + Na_2CO_3(aq) \rightarrow CuCO_3(s) + 2NaNO_3(aq)$

c. $CoCl_2(aq) + 2NaOH(aq) \rightarrow Co(OH)_2(s) + 2NaCl(aq)$

$SnCl_2(aq) + 2NaOH(aq) \rightarrow Sn(OH)_2(s) + 2NaCl(aq)$

$Cu(NO_3)_2(aq) + 2NaOH(aq) \rightarrow Cu(OH)_2(s) + 2NaNO_3(aq)$

d. $3CoCl_2(aq) + 2Na_3PO_4(aq) \rightarrow Co_3(PO_4)_2(s) + 6NaCl(aq)$

$3SnCl_2(aq) + 2Na_3PO_4(aq) \rightarrow Sn_3(PO_4)_2(s) + 6NaCl(aq)$

$3Cu(NO_3)_2(aq) + 2Na_3PO_4(aq) \rightarrow Cu_3(PO_4)_2(s) + 6NaNO_3(aq)$

Chapter 8   Classifying Chemical Reactions

2. transfer

4. Metal atoms *lose* electrons and form *positive ions* (cations); nonmetal atoms *gain* electrons and become *negative ions* (anions).

6. Bromine atoms each gain one electron to form bromide ions (which have a 1− charge). Since the element bromine occurs naturally as Br_2 molecules, each bromine molecule would gain two electrons, forming two bromide ions, each with a 1− charge.

8. $AlBr_3$ is made up of Al^{3+} ions and Br^- ions. Aluminum atoms each lose three electrons to become Al^{3+} ions. Bromine atoms each gain one electron to become Br^- ions (so each Br_2 molecule gains two electrons to become two Br^- ions).

10.
    a. $Na(s) + S(s) \rightarrow Na_2S(s)$
          Balance sodium: $2Na(s) + S(s) \rightarrow Na_2S(s)$
          Balanced equation: $2Na(s) + S(s) \rightarrow Na_2S(s)$

    b. $Cu(s) + AgNO_3(aq) \rightarrow Cu(NO_3)_2(aq) + Ag(s)$
          Balance nitrate: $Cu(s) + 2AgNO_3(aq) \rightarrow Cu(NO_3)_2(aq) + Ag(s)$
          Balance silver: $Cu(s) + AgNO_3(aq) \rightarrow Cu(NO_3)_2(aq) + 2Ag(s)$
          Balanced equation: $Cu(s) + 2AgNO_3(aq) \rightarrow Cu(NO_3)_2(aq) + 2Ag(s)$

    c. $Mg(s) + HCl(aq) \rightarrow MgCl_2(aq) + H_2(g)$
          Balance chlorine: $Mg(s) + 2HCl(aq) \rightarrow MgCl_2(aq) + H_2(g)$
          Balanced equation: $Mg(s) + 2HCl(aq) \rightarrow MgCl_2(aq) + H_2(g)$

    d. $Pb(s) + S(s) \rightarrow PbS_2(s)$
          Balance sulfur: $Pb(s) + 2S(s) \rightarrow PbS_2(s)$
          Balanced equation: $Pb(s) + 2S(s) \rightarrow PbS_2(s)$

    e. $Li(s) + H_2O(l) \rightarrow LiOH(aq) + H_2(g)$
          Balance hydrogen: $Li(s) + 2H_2O(l) \rightarrow 2LiOH(aq) + H_2(g)$
          Balance lithium: $2Li(s) + 2H_2O(l) \rightarrow 2LiOH(aq) + H_2(g)$
          $2Li(s) + 2H_2O(l) \rightarrow 2LiOH(aq) + H_2(g)$

12. examples of formation of water:
        $HCl(aq) + NaOH(aq) \rightarrow H_2O(l) + NaCl(aq)$
        $H_2SO_4(aq) + 2KOH(aq) \rightarrow 2H_2O(l) + K_2SO_4(aq)$
    examples of formation of a gaseous product
        $Mg(s) + 2HCl(aq) \rightarrow MgCl_2(aq) + H_2(g)$
        $2KClO_3(s) \rightarrow 2KCl(s) + 3O_2(g)$

64   Chapter 8   Classifying Chemical Reactions

14. For each reaction, the type of reaction is first identified, followed by some of the reasoning that leads to this choice (there may be more than one way in which you can recognize a particular type of reaction).

   a. oxidation-reduction (oxygen changes from the combined state to the elemental state)

   b. oxidation-reduction (copper changes from the elemental to the combined state; hydrogen changes from the combined to the elemental state)

   c. acid-base (H_2SO_4 is a strong acid and NaOH is a strong base; water and a salt are formed)

   d. acid-base, precipitation (H_2SO_4 is a strong acid, and $Ba(OH)_2$ is a base; water and a salt are formed; an insoluble product forms)

   e. precipitation (from the Solubility Rules of Table 7.1, AgCl is only slightly soluble)

   f. precipitation (from the Solubility Rules of Table 7.1, $Cu(OH)_2$ is only slightly soluble)

   g. oxidation-reduction (chlorine and fluorine change from the elemental to the combined state)

   h. oxidation-reduction (oxygen changes from the elemental to the combined state)

   i. acid-base (HNO_3 is a strong acid and $Ca(OH)_2$ is a strong base; a salt and water are formed)

16. oxidation-reduction

18. A decomposition reaction is one in which a given compound is broken down into simpler compounds or constituent elements. The reactions

   $$CaCO_3(s) \rightarrow CaO(s) + CO_2(g)$$

   $$2HgO(s) \rightarrow 2Hg(l) + O_2(g)$$

   both represent decomposition reactions. Such reactions often (but not necessarily always) may be classified in other ways. For example, the reaction of HgO(s) is also an oxidation-reduction reaction.

20. Compounds like those in this problem, containing only carbon and hydrogen, are called *hydrocarbons*. When a hydrocarbon is reacted with oxygen (O_2), the hydrocarbon is almost always converted to carbon dioxide and water vapor. Since water molecules contain an odd number of oxygen atoms, whereas O_2 contains an even number of oxygen atoms, it is often difficult to balance such equations. In particular, if the hydrocarbon being reacted has an *even number of carbon atoms*, the balancing can really become complicated. A trick that often simplifies the balancing

is, if the hydrocarbon contains an *even* number of carbon atoms, put a coefficient of *two* in front of the hydrocarbon before continuing to balance.

    a. $C_3H_8(l) + O_2(g) \rightarrow CO_2(g) + H_2O(g)$

        Balance carbon: $C_3H_8(l) + O_2(g) \rightarrow 3CO_2(g) + H_2O(g)$

        Balance hydrogen: $C_3H_8(l) + O_2(g) \rightarrow 3CO_2(g) + 4H_2O(g)$

        Balance oxygen: $C_3H_8(l) + 5O_2(g) \rightarrow 3CO_2(g) + 4H_2O(g)$

        Balanced equation: $C_3H_8(l) + 5O_2(g) \rightarrow 3CO_2(g) + 4H_2O(g)$

    b. $C_4H_{10}(l) + O_2(g) \rightarrow CO_2(g) + H_2O(g)$

        Apply the trick above: $2C_4H_{10}(l) + O_2(g) \rightarrow CO_2(g) + H_2O(g)$

        Balance carbon: $2C_4H_{10}(l) + O_2(g) \rightarrow 8CO_2(g) + H_2O(g)$

        Balance hydrogen: $2C_4H_{10}(l) + O_2(g) \rightarrow 8CO_2(g) + 10H_2O(g)$

        Balance oxygen: $2C_4H_{10}(l) + 13O_2(g) \rightarrow 8CO_2(g) + 10H_2O(g)$

        Balanced equation: $2C_4H_{10}(l) + 13O_2(g) \rightarrow 8CO_2(g) + 10H_2O(g)$

    c. $C_5H_{12}(l) + O_2(g) \rightarrow CO_2(g) + H_2O(g)$

        Balance carbon: $C_5H_{12}(l) + O_2(g) \rightarrow 5CO_2(g) + H_2O(g)$

        Balance hydrogen: $C_5H_{12}(l) + O_2(g) \rightarrow 5CO_2(g) + 6H_2O(g)$

        Balance oxygen: $C_5H_{12}(l) + 8O_2(g) \rightarrow 5CO_2(g) + 6H_2O(g)$

        Balanced equation: $C_5H_{12}(l) + 8O_2(g) \rightarrow 5CO_2(g) + 6H_2O(g)$

22.   a. $C_{19}H_{40}(s) + O_2(g) \rightarrow CO_2(g) + H_2O(g)$

        Balance carbon: $C_{19}H_{40}(s) + O_2(g) \rightarrow 19CO_2(g) + H_2O(g)$

        Balance hydrogen: $C_{19}H_{40}(s) + O_2(g) \rightarrow 19CO_2(g) + 20H_2O(g)$

        Balance oxygen: $C_{19}H_{40}(s) + 29O_2(g) \rightarrow 19CO_2(g) + 20H_2O(g)$

        Balanced equation: $C_{19}H_{40}(s) + 29O_2(g) \rightarrow 19CO_2(g) + 20H_2O(g)$

    b. $C_6H_{12}O_6(s) + O_2(g) \rightarrow CO_2(g) + H_2O(g)$

        Balance carbon: $C_6H_{12}O_6(s) + O_2(g) \rightarrow 6CO_2(g) + H_2O(g)$

        Balance hydrogen: $C_6H_{12}O_6(s) + O_2(g) \rightarrow 6CO_2(g) + 6H_2O(g)$

        Balance oxygen: $C_6H_{12}O_6(s) + 6O_2(g) \rightarrow 6CO_2(g) + 6H_2O(g)$

        Balanced equation: $C_6H_{12}O_6(s) + 6O_2(g) \rightarrow 6CO_2(g) + 6H_2O(g)$

    c. $C_{12}H_{22}O_{11}(s) + 12O_2(g) \rightarrow 12CO_2(g) + 11H_2O(g)$

        Balance carbon: $C_{12}H_{22}O_{11}(s) + O_2(g) \rightarrow 12CO_2(g) + H_2O(g)$

        Balance hydrogen: $C_{12}H_{22}O_{11}(s) + O_2(g) \rightarrow 12CO_2(g) + 11H_2O(g)$

Balance oxygen: $C_{12}H_{22}O_{11}(s) + 12O_2(g) \rightarrow 12CO_2(g) + 11H_2O(g)$

Balanced equation: $C_{12}H_{22}O_{11}(s) + 12O_2(g) \rightarrow 12CO_2(g) + 11H_2O(g)$

24. a. $2Co(s) + 3S(s) \rightarrow Co_2S_3(s)$
    b. $2NO(g) + O_2(g) \rightarrow 2NO_2(g)$
    c. $FeO(s) + CO_2(g) \rightarrow FeCO_3(s)$
    d. $2Al(s) + 3F_2(g) \rightarrow 2AlF_3(s)$
    e. $2NH_3(g) + H_2CO_3(aq) \rightarrow (NH_4)_2CO_3(s)$

26. a. $2NI_3(s) \rightarrow N_2(g) + 3I_2(s)$
    b. $BaCO_3(s) \rightarrow BaO(s) + CO_2(g)$
    c. $C_6H_{12}O_6(s) \rightarrow 6C(s) + 6H_2O(g)$
    d. $Cu(NH_3)_4SO_4(s) \rightarrow CuSO_4(s) + 4NH_3(g)$
    e. $3NaN_3(s) \rightarrow Na_3N(s) + 4N_2(g)$

28. a. two; $O + 2e^- \rightarrow O^{2-}$
    b. one; $F + e^- \rightarrow F^-$
    c. three; $N + 3e^- \rightarrow N^{3-}$
    d. one; $Cl + e^- \rightarrow Cl^-$
    e. two; $S + 2e^- \rightarrow S^{2-}$
    f. one; $Br + e^- \rightarrow Br^-$
    g. one; $I + e^- \rightarrow I^-$

30. carbon dioxide gas and water vapor

32. a. $2C_3H_8O(l) + 9O_2(g) \rightarrow 6CO_2(g) + 8H_2O(g)$
       oxidation-reduction, combustion
    b. $HCl(aq) + AgC_2H_3O_2(aq) \rightarrow AgCl(s) + HC_2H_3O_2(aq)$
       precipitation, double-displacement
    c. $3HCl(aq) + Al(OH)_3(s) \rightarrow AlCl_3(aq) + 3H_2O(l)$
       acid-base, double-displacement
    d. $2H_2O_2(aq) \rightarrow 2H_2O(l) + O_2(g)$
       oxidation-reduction, decomposition
    e. $N_2H_4(l) + O_2(g) \rightarrow N_2(g) + 2H_2O(g)$
       oxidation-reduction, combustion

34. $2Zn(s) + O_2(g) \rightarrow 2ZnO(s)$

$4Al(s) + 3O_2(g) \rightarrow 2Al_2O_3(s)$

$2Fe(s) + O_2(g) \rightarrow 2FeO(s)$; $4Fe(s) + 3O_2(g) \rightarrow 2Fe_2O_3(s)$

$2Cr(s) + O_2(g) \rightarrow 2CrO(s)$; $4Cr(s) + 3O_2(g) \rightarrow 2Cr_2O_3(s)$

$2Ni(s) + O_2(g) \rightarrow 2NiO(s)$

36.
Al	3+		Ba	2+
Br	1−		Ca	2+
Cl	1−		Cs	1+
I	1−		K	1+
Li	1+		Mg	2+
Na	1+		O	2−
Rb	1+		S	2−
Sr	2+			

38. K^+, Ca^{2+}

40. a. $Na + O_2 \rightarrow Na_2O_2$

Balance sodium: $2Na + O_2 \rightarrow Na_2O_2$

Balanced equation: $2Na(s) + O_2(g) \rightarrow Na_2O_2(s)$

b. $Fe(s) + H_2SO_4(aq) \rightarrow FeSO_4(aq) + H_2(g)$

Equation is already balanced!

c. $Al_2O_3 \rightarrow Al + O_2$

Balance oxygen: $2Al_2O_3 \rightarrow Al + 3O_2$

Balance aluminum: $2Al_2O_3 \rightarrow 4Al + 3O_2$

Balanced equation: $2Al_2O_3(s) \rightarrow 4Al(s) + 3O_2(g)$

d. $Fe + Br_2 \rightarrow 2FeBr_3$

Balance bromine: $Fe + 3Br_2 \rightarrow 2FeBr_3$

Balance iron: $2Fe + 3Br_2 \rightarrow 2FeBr_3$

Balanced equation: $2Fe(s) + 3Br_2(l) \rightarrow 2FeBr_3(s)$

e. $Zn + HNO_3 \rightarrow Zn(NO_3)_2 + H_2$

Balance nitrate ions: $Zn + 2HNO_3 \rightarrow Zn(NO_3)_2 + H_2$

Balanced equation: $Zn(s) + 2HNO_3(aq) \rightarrow Zn(NO_3)_2(aq) + H_2(g)$

42. a. $C_5H_{12}(l) + 8O_2(g) \rightarrow 5CO_2(g) + 6H_2O(g)$
    b. $C_2H_6O(l) + 3O_2(g) \rightarrow 2CO_2(g) + 3H_2O(g)$
    c. $2C_6H_6(l) + 15O_2(g) \rightarrow 12CO_2(g) + 6H_2O(g)$

44. a. $4FeO(s) + O_2(g) \rightarrow 2Fe_2O_3(s)$
    b. $2CO(g) + O_2(g) \rightarrow 2CO_2(g)$
    c. $H_2(g) + Cl_2(g) \rightarrow 2HCl(g)$
    d. $16K(s) + S_8(s) \rightarrow 8K_2S(s)$
    e. $6Na(s) + N_2(g) \rightarrow 2Na_3N(s)$

46. For simplicity, the physical states of the substances are omitted.

    $2Ba + O_2 \rightarrow 2BaO$ $Ba + S \rightarrow BaS$
    $Ba + Cl_2 \rightarrow BaCl_2$ $3Ba + N_2 \rightarrow Ba_3N_2$
    $Ba + Br_2 \rightarrow BaBr_2$ $4K + O_2 \rightarrow 2K_2O$
    $2K + S \rightarrow K_2S$ $2K + Cl_2 \rightarrow 2KCl$
    $6K + N_2 \rightarrow 2K_3N$ $2K + Br_2 \rightarrow 2KBr$
    $2Mg + O_2 \rightarrow 2MgO$ $Mg + S \rightarrow MgS$
    $Mg + Cl_2 \rightarrow MgCl_2$ $3Mg + N_2 \rightarrow Mg_3N_2$
    $Mg + Br_2 \rightarrow MgBr_2$ $4Rb + O_2 \rightarrow 2Rb_2O$
    $2Rb + S \rightarrow Rb_2S$ $2Rb + Cl_2 \rightarrow 2RbCl$
    $6Rb + N_2 \rightarrow 2Rb_3N$ $2Rb + Br_2 \rightarrow 2RbBr$
    $2Ca + O_2 \rightarrow 2CaO$ $Ca + S \rightarrow CaS$
    $Ca + Cl_2 \rightarrow CaCl_2$ $3Ca + N_2 \rightarrow Ca_3N_2$
    $Ca + Br_2 \rightarrow CaBr_2$ $4Li + O_2 \rightarrow 2Li_2O$
    $2Li + S \rightarrow Li_2S$ $2Li + Cl_2 \rightarrow 2LiCl$
    $6Li + N_2 \rightarrow 2Li_3N$ $2Li + Br_2 \rightarrow 2LiBr$

48. $Fe(s) + H_2SO_4(aq) \rightarrow FeSO_4(aq) + H_2(g)$
    $Zn(s) + H_2SO_4(aq) \rightarrow ZnSO_4(aq) + H_2(g)$
    $Mg(s) + H_2SO_4(aq) \rightarrow MgSO_4(aq) + H_2(g)$
    $Co(s) + H_2SO_4(aq) \rightarrow CoSO_4(aq) + H_2(g)$
    $Ni(s) + H_2SO_4(aq) \rightarrow NiSO_4(aq) + H_2(g)$

50. $16Na + S_8 \rightarrow 8Na_2S$ $16K + S_8 \rightarrow 8K_2S$ $8Ba + S_8 \rightarrow 8BaS$
    $8Ca + S_8 \rightarrow 8CaS$ $16Al + 3S_8 \rightarrow 8Al_2S_3$ $8Mg + S_8 \rightarrow 8MgS$
    $8Fe + S_8 \rightarrow 8FeS$

Cumulative Review: Chapters 6, 7, and 8

2. A chemical equation indicates the substances necessary for a chemical reaction to take place, as well as what is produced by that chemical reaction. The substances to the left of the arrow in a chemical equation are called the reactants; those to the right of the arrow are referred to as the products. In addition, if a chemical equation has been balanced, then the equation indicates the relative proportions in which the reactant molecules combine to form the product molecules.

4. It is *never* permissible to change the subscripts of a formula when balancing a chemical equation: changing the subscripts changes the *identity* of a substance from one chemical to another. For example, consider the unbalanced chemical equation

$$H_2(g) + O_2(g) \rightarrow H_2O(l)$$

If you changed the *formula* of the product from $H_2O(l)$ to $H_2O_2(l)$, the equation would appear to be "balanced". However, H_2O is water, whereas H_2O_2 is hydrogen peroxide--a completely different chemical substance (which is not prepared by reaction of the elements hydrogen and oxygen).

When we balance a chemical equation, it is permitted only to adjust the *coefficients* of a formula, since changing a coefficient merely changes the number of molecules of a substance being used in the reaction, without changing the identity of the substance. For the example above, we can balance the equation by putting coefficients of 2 in front of the formulas of H_2 and H_2O: these coefficients do not change the nature of what is reacting and what product is formed.

$$2H_2(g) + O_2(g) \rightarrow 2H_2O(l)$$

6. A precipitation reaction is one in which a *solid* forms when the reactants are combined: the solid is called a precipitate. If you were to perform such a reaction, the mixture would turn cloudy as the reactants are combined, and a solid would eventually settle from the mixture on standing. There are many examples of such precipitation reactions: consult the solubility rules in Table 7.1 if you need help. One example would be to combine barium nitrate and sodium carbonate solutions: a precipitate of barium carbonate would form.

$$Ba(NO_3)_2(aq) + Na_2CO_3(aq) \rightarrow BaCO_3(s) + 2NaNO_3(aq)$$

8. In summary, nearly all compounds containing the nitrate, sodium, potassium, and ammonium ions are soluble in water. Most salts containing the chloride and sulfate ions are soluble in water, with specific exceptions (see Table 7.1 for these exceptions). Most compounds containing the hydroxide, sulfide, carbonate, and phosphate ions are not soluble in water, unless the compound also contains one of the cations mentioned above (Na^+, K^+, NH_4^+).

The solubility rules are phrased as if you had a sample of a given solute and wanted to see if you could dissolve it in water. These rules can also be applied, however, to predict the identity of the solid produced in a precipitation reaction: a given combination of ions will not be soluble in water whether you take a pure compound out of a reagent bottle or if you generate the insoluble combination of ions during a chemical reaction. For example, the solubility rules say that BaSO4 is not soluble in water. This means not only that a pure sample of BaSO4 taken from a reagent bottle will not dissolve in water, but also that if Ba^{2+} ion and SO_4^{2-} ion end up together in the same solution, they will precipitate as $BaSO_4$. If we were to combine barium chloride and sulfuric acid solutions

$$BaCl_2(aq) + H_2SO_4(aq) \rightarrow BaSO_4(s) + 2HCl(aq)$$

then, because barium sulfate is not soluble in water, a precipitate of $BaSO_4(s)$ would form. Since a precipitate of $BaSO_4(s)$ would form no matter what barium compound or what sulfate compound were mixed, we can write the net ionic equation for the reaction as

$$Ba^{2+}(aq) + SO_4^{2-}(aq) \rightarrow BaSO_4(s)$$

Thus if, for example, barium nitrate solution were combined with sodium sulfate solution, a precipitate of $BaSO_4$ would form. Barium sulfate is insoluble in water regardless of its source.

10. Acids (such as the citric acid found in citrus fruits and the acetic acid found in vinegar) were first noted primarily because of their sour taste. The first bases noted were characterized by their bitter taste and slippery feel on the skin. Acids and bases chemically react with (neutralize) each other forming water: the net ionic equation is

$$H^+(aq) + OH^-(aq) \rightarrow H_2O(l)$$

The *strong* acids and bases are those which fully ionize when they dissolve in water: since these substances fully ionize, they are strong electrolytes. The common strong acids are HCl(hydrochloric), HNO_3(nitric), H_2SO_4(sulfuric), and $HClO_4$(perchloric). The most common strong bases are the alkali metal hydroxides, particularly NaOH(sodium hydroxide) and KOH(potassium hydroxide).

12. Oxidation-reduction reactions are electron-transfer reactions. Oxidation represents a loss of electrons by an atom, molecule, or ion, whereas reduction is the gain of electrons by such a species. Since an oxidation-reduction process represents the transfer of electrons between species, you can't have one without the other also taking place: the electrons lost by one species must be gained by some other species. An example of a simple oxidation reduction reaction between a metal and a nonmetal could be the following

$$Mg(s) + F_2(g) \rightarrow MgF_2(s)$$

In this process, Mg atoms lose two electrons each to become Mg^{2+} ions in MgF_2: Mg is oxidized. Each F atom of F_2 gains one electron to become an F^- ion, for a total of two electrons gained for each F_2 molecule: F_2 is reduced.

$$Mg \rightarrow Mg^{2+} + 2e^- \qquad 2(F + e^- \rightarrow F^-)$$

14. In general, a synthesis reaction represents the reaction of elements or simple compounds to produce more complex substances. There are many examples of synthesis reactions, for example

$$N_2(g) + 3H_2(g) \rightarrow 2NH_3(g)$$

$$NaOH(aq) + CO_2(g) \rightarrow NaHCO_3(s)$$

Decomposition reactions represent the breakdown of a more complex substance into simpler substances. There are many examples of decomposition reactions, for example

$$2H_2O_2(aq) \rightarrow 2H_2O(l) + O_2(g)$$

Synthesis and decomposition reactions are very often also oxidation-reduction reactions, especially if an elemental substance reacts or is generated. It is not necessary, however, for synthesis and decomposition reactions to always involve oxidation-reduction. For example, the reaction between NaOH and CO_2 given as an example of a synthesis reaction does *not* represent oxidation-reduction.

16. a. $2Na(s) + 2H_2O(l) \rightarrow 2NaOH(aq) + H_2(g)$

    $Ca(s) + 2H_2O(l) \rightarrow Ca(OH)_2(aq) + H_2(g)$

    b. $2Na(s) + Br_2(l) \rightarrow 2NaBr(s)$

    $Ca(s) + Br_2(l) \rightarrow CaBr_2(s)$

    c. $6Na(s) + N_2(g) \rightarrow 2Na_3N(s)$

    $3Ca(s) + N_2(g) \rightarrow Ca_3N_2(s)$

    d. $2Na(s) + O_2(g) \rightarrow Na_2O_2(s)$

    $2Ca(s) + O_2(g) \rightarrow 2CaO(s)$

    e. $2Na(s) + S(s) \rightarrow Na_2S(s)$

    $Ca(s) + S(s) \rightarrow CaS(s)$

18. a. $Ba(NO_3)_2(aq) + K_2CrO_4(aq) \rightarrow BaCrO_4(s) + 2KNO_3(aq)$

    b. $NaOH(aq) + CH_3COOH(aq) \rightarrow H_2O(l) + NaCH_3COO(aq)$
        (then evaporate the water from the solution)

    c. $AgNO_3(aq) + NaCl(aq) \rightarrow AgCl(s) + NaNO_3(aq)$

    d. $Pb(NO_3)_2(aq) + H_2SO_4(aq) \rightarrow PbSO_4(s) + 2HNO_3(aq)$

e. $2NaOH(aq) + H_2SO_4(aq) \rightarrow Na_2SO_4(aq) + 2H_2O(l)$
(then evaporate the water from the solution)

f. $Ba(NO_3)_2(aq) + Na_2CO_3(aq) \rightarrow BaCO_3(s) + 2NaNO_3(aq)$

20. a. $FeO(s) + 2HNO_3(aq) \rightarrow Fe(NO_3)_2(aq) + H_2O(l)$
acid-base, double-displacement

b. $2Mg(s) + 2CO_2(g) + O_2(g) \rightarrow 2MgCO_3(s)$
synthesis; oxidation-reduction

c. $2NaOH(s) + CuSO_4(aq) \rightarrow Cu(OH)_2(s) + Na_2SO_4(aq)$
precipitation, double-displacement

d. $HI(aq) + KOH(aq) \rightarrow KI(aq) + H_2O(l)$
acid-base, double-displacement

e. $C_3H_8(g) + 5O_2(g) \rightarrow 3CO_2(g) + 4H_2O(g)$
combustion; oxidation-reduction

f. $Co(NH_3)_6Cl_2(s) \rightarrow CoCl_2(s) + 6NH_3(g)$
decomposition

g. $2HCl(aq) + Pb(C_2H_3O_2)_2(aq) \rightarrow 2HC_2H_3O_2(aq) + PbCl_2(s)$
precipitation, double-displacement

h. $C_{12}H_{22}O_{11}(s) \rightarrow 12C(s) + 11H_2O(g)$
decomposition; oxidation-reduction

i. $2Al(s) + 6HNO_3(aq) \rightarrow 2Al(NO_3)_3(aq) + 3H_2(g)$
oxidation-reduction

j. $4B(s) + 3O_2(g) \rightarrow 2B_2O_3(s)$
synthesis; oxidation-reduction

Chapter 9   Chemical Composition

2. $500. \text{ g} \times \dfrac{1 \text{ cork}}{1.63 \text{ g}} = 306.7 = 307 \text{ corks}$

    $500. \text{ g} \times \dfrac{1 \text{ stopper}}{4.31 \text{ g}} = 116 \text{ stoppers}$

    1 kg (1000 g) of corks contains $(1000 \text{ g} \times \dfrac{1 \text{ cork}}{1.63 \text{ g}}) = 613.49 = 613 \text{ corks}$

    613 stoppers would weigh $(613 \text{ stoppers} \times \dfrac{4.31 \text{ g}}{1 \text{ stopper}}) = 2644 \text{ g} = 2640 \text{ g}$

    The ratio of the mass of a stopper to the mass of a cork is (4.31 g/1.63 g). So the mass of stoppers that contains the same number of stoppers as there are corks in 1000 g of corks is

    $1000 \text{ g} \times \dfrac{4.31 \text{ g}}{1.63 \text{ g}} = 2644 \text{ g} = 2640 \text{ g}$

4. the same

6. a. $10.81 \text{ amu} \times \dfrac{1 \text{ B atom}}{10.81 \text{ amu}} = 1.000 \text{ atom} = 1 \text{ B atom}$

    b. $320.7 \text{ amu} \times \dfrac{1 \text{ S atom}}{32.07 \text{ amu}} = 10 \text{ S atoms}$

    c. $19{,}691 \text{ amu} \times \dfrac{1 \text{ Au atom}}{197.0 \text{ amu}} = 99.95 \text{ Au atoms} = 100 \text{ Au atoms}$

    d. $19{,}695 \text{ amu} \times \dfrac{1 \text{ Xe atom}}{131.3 \text{ amu}} = 150.0 \text{ Xe atoms} = 150 \text{ Xe atoms}$

    e. $3588.3 \text{ amu} \times \dfrac{1 \text{ Al atom}}{26.98 \text{ amu}} = 133.0 \text{ Al atoms} = 133 \text{ Al atoms}$

8. $1.00 \times 10^4 \text{ atoms} \times \dfrac{196.97 \text{ amu}}{1 \text{ Au atom}} = 1.97 \times 10^6 \text{ amu}$

    $2.955 \times 10^5 \text{ amu} \times \dfrac{1 \text{ Au atom}}{196.97 \text{ amu}} = 1500. \text{ Au atoms}$

10. Avogadro's number (6.022×10^{23})

12. The ratio of the atomic mass of C to the atomic mass of H is (12.01 amu/1.008 amu), and the mass of carbon is given by

    $1.008 \text{ g} \times \dfrac{12.01 \text{ amu}}{1.008 \text{ amu}} = 12.01 \text{ g C}$

74   Chapter 9   Chemical Composition

14. The ratio of the atomic mass of N to the atomic mass of O is (14.01 amu/16.00 amu), and the mass of nitrogen is given by

$$48 \text{ g} \times \frac{14.01 \text{ amu}}{16.00 \text{ amu}} = 42 \text{ g N}$$

16. 1 mol O = 16.00 g O = 6.02 × 10^{23} O atoms

$$1 \text{ O atom} \times \frac{16.00 \text{ g O}}{6.022 \times 10^{23} \text{ O atoms}} = 2.66 \times 10^{-23} \text{ g}$$

18. $0.50 \text{ mol O atoms} \times \dfrac{16.00 \text{ g}}{1 \text{ mol}} = 8.0 \text{ g O}$

$4 \text{ mol H atoms} \times \dfrac{1.008 \text{ g}}{1 \text{ mol}} = 4 \text{ g H}$

Half a mole of O atoms weighs more than 4 moles of H atoms.

20. a. $26.2 \text{ g Au} \times \dfrac{1 \text{ mol Au}}{197.0 \text{ g}} = 0.133 \text{ mol Au}$

b. $41.5 \text{ g Ca} \times \dfrac{1 \text{ mol Ca}}{40.08 \text{ g}} = 1.04 \text{ mol Ca}$

c. $335 \text{ mg Ba} \times \dfrac{1 \text{ g}}{10^3 \text{ mg}} \times \dfrac{1 \text{ mol Ba}}{137.3 \text{ g}} = 2.44 \times 10^{-3} \text{ mol Ba}$

d. $1.42 \times 10^{-3} \text{ g Pd} \times \dfrac{1 \text{ mol Pd}}{106.4 \text{ g}} = 1.33 \times 10^{-5} \text{ mol Pd}$

e. $3.05 \times 10^{-5} \text{ μg Ni} \times \dfrac{1 \text{ g}}{10^6 \text{ μg}} \times \dfrac{1 \text{ mol Ni}}{58.70 \text{ g}} = 5.20 \times 10^{-13} \text{ mol Ni}$

f. $1.00 \text{ lb Fe} \times \dfrac{453.59 \text{ g}}{1 \text{ lb}} \times \dfrac{1 \text{ mol Fe}}{55.85 \text{ g}} = 8.12 \text{ mol Fe}$

g. $12.01 \text{ g C} \times \dfrac{1 \text{ mol C}}{12.01 \text{ g}} = 1.00 \text{ mol C}$

22. a. $2.00 \text{ mol Fe} \times \dfrac{55.85 \text{ g}}{1 \text{ mol}} = 112 \text{ g Fe}$

b. $0.521 \text{ mol Ni} \times \dfrac{58.70 \text{ g}}{1 \text{ mol}} = 30.6 \text{ g Ni}$

c. $1.23 \times 10^{-3} \text{ mol Pt} \times \dfrac{195.1 \text{ g}}{1 \text{ mol}} = 0.240 \text{ g Pt}$

d. $72.5 \text{ mol Pb} \times \dfrac{207.2 \text{ g}}{1 \text{ mol}} = 1.50 \times 10^4 \text{ g Pb}$

e. $0.00102 \text{ mol Mg} \times \dfrac{24.31 \text{ g}}{1 \text{ mol}} = 0.0248 \text{ g Mg}$

f. $4.87 \times 10^3 \text{ mol Al} \times \dfrac{26.98 \text{ g}}{1 \text{ mol}} = 1.31 \times 10^5 \text{ g Al}$

g. $211.5 \text{ mol Li} \times \dfrac{6.941 \text{ g}}{1 \text{ mol}} = 1468 \text{ g Li}$

h. $1.72 \times 10^{-6} \text{ mol Na} \times \dfrac{22.99 \text{ g}}{1 \text{ mol}} = 3.95 \times 10^{-5} \text{ g Na}$

24. a. $6.42 \times 10^{22} \text{ atoms Li} \times \dfrac{6.941 \text{ g Li}}{6.022 \times 10^{23} \text{ atoms Li}} = 0.740 \text{ g Li}$

b. $0.551 \text{ mol Li} \times \dfrac{6.941 \text{ g Li}}{1 \text{ mol Li}} = 3.82 \text{ g Li}$

c. $0.00103 \text{ mol Li} \times \dfrac{6.022 \times 10^{23} \text{ atoms Li}}{1 \text{ mol Li}} = 6.20 \times 10^{20} \text{ Li atoms}$

d. $1 \text{ Li atom} \times \dfrac{6.941 \text{ g Li}}{6.022 \times 10^{23} \text{ Li atoms}} = 1.153 \times 10^{-23} \text{ g}$

e. $0.500 \text{ g} \times \dfrac{1 \text{ mol Li}}{6.941 \text{ g Li}} = 0.0720 \text{ mol Li}$

f. $5.22 \text{ g Li} \times \dfrac{6.022 \times 10^{23} \text{ Li atoms}}{6.941 \text{ g Li}} = 4.53 \times 10^{23} \text{ Li atoms}$

g. $4.75 \times 10^{24} \text{ Li atoms} \times \dfrac{1 \text{ mol Li}}{6.022 \times 10^{23} \text{ Li atoms}} = 7.89 \text{ mol Li}$

26. adding together (summing)

28. a. mass of 3 mol Na = 3(22.99 g) = 68.97 g
    mass of 1 mol N = 1(14.01 g) = 14.01 g
    molar mass of Na$_3$N = 82.98 g

b. mass of 1 mol C = 12.01 g = 12.01 g
    mass of 2 mol S = 2(32.07 g) = 64.14 g
    molar mass of CS$_2$ = 76.15 g

c. | mass of 1 mol N = | 14.01 g = | 14.01 g
   | mass of 4 mol H = | 4(1.008 g) = | 4.032 g
   | mass of 1 mol Br = | 79.90 g = | 79.90 g
   | molar mass of NH_4Br = | | 97.942 g = 97.94 g

d. | mass of 2 mol C = | 2(12.01 g) = | 24.02 g
   | mass of 6 mol H = | 6(1.008 g) = | 6.048 g
   | mass of 1 mol O = | 16.00 g = | 16.00 g
   | molar mass of C_2H_6O = | | 46.07 g

e. | mass of 2 mol H = | 2(1.008 g) = | 2.016 g
   | mass of 1 mol S = | 32.07 g = | 32.07 g
   | mass of 3 mol O = | 3(16.00 g) = | 48.00 g
   | molar mass of H_2SO_3 = | | 82.086 g = 82.09 g

f. | mass of 2 mol H = | 2(1.008 g) = | 2.016 g
   | mass of 1 mol S = | 32.07 g = | 32.07 g
   | mass of 4 mol O = | 4(16.00 g) = | 64.00 g
   | molar mass of H_2SO_4 = | | 98.086 g = 98.09 g

30. a. | mass of 2 mol N = | 2(14.01 g) = | 28.02 g
   | mass of 8 mol H = | 8(1.008 g) = | 8.064 g
   | mass of 1 mol S = | 32.07 g = | 32.07 g
   | molar mass of $(NH_4)_2S$ = | | 68.154 g = 68.15 g

b. | mass of 6 mol C = | 6(12.01 g) = | 72.06 g
   | mass of 4 mol H = | 4(1.008 g) = | 4.032 g
   | mass of 1 mol O = | 16.00 g = | 16.00 g
   | mass of 2 mol Cl = | 2(35.45 g) = | 70.90 g
   | molar mass of $C_6H_4OCl_2$ = | | 162.992 g = 162.99 g

c. | mass of 1 mol Ba = | 137.34 g = | 137.34 g
| mass of 2 mol H = | 2(1.008 g) = | 2.016 g
| molar mass of BaH$_2$ = | | 139.356 g = 139.36 g

d. | mass of 1 mol K = | 39.10 g = | 39.10 g
| mass of 2 mol H = | 2(1.008 g) = | 2.016 g
| mass of 1 mol P = | 30.97 g = | 30.97 g
| mass of 4 mol O = | 4(16.00 g) = | 64.00 g
| molar mass of KH$_2$PO$_4$ = | | 136.086 g = 136.09 g

e. | mass of 2 mol K = | 2(39.10 g) = | 78.20 g
| mass of 1 mol H = | 1.008 g = | 1.008 g
| mass of 1 mol P = | 30.97 g = | 30.97 g
| mass of 4 mol O = | 4(16.00 g) = | 64.00 g
| molar mass of K$_2$HPO$_4$ = | | 174.178 g = 174.18 g

f. | mass of 3 mol K = | 3(39.10 g) = | 117.3 g
| mass of 1 mol P = | 30.97 g = | 30.97 g
| mass of 4 mol O = | 4(16.00 g) = | 64.00 g
| molar mass of K$_3$PO$_4$ = | | 212.27 g = 212.3 g

32. a. molar mass of SO$_3$ = 80.07 g

$$49.2 \text{ mg SO}_3 \times \frac{1 \text{ g}}{1000 \text{ mg}} \times \frac{1 \text{ mol}}{80.07 \text{ g}} = 6.14 \times 10^{-4} \text{ mol SO}_3$$

b. molar mass of PbO$_2$ = 239.2 g

$$7.44 \times 10^4 \text{ kg PbO}_2 \times \frac{1000 \text{ g}}{1 \text{ kg}} \times \frac{1 \text{ mol}}{239.2 \text{ g}} = 3.11 \times 10^5 \text{ mol PbO}_2$$

c. molar mass of CHCl$_3$ 119.37 g

$$59.1 \text{ g CHCl}_3 \times \frac{1 \text{ mol}}{119.37 \text{ g}} = 0.495 \text{ mol CHCl}_3$$

78 Chapter 9 Chemical Composition

    d.   molar mass of $C_2H_3Cl_3$ = 133.39 g

$$3.27 \ \mu g \times \frac{1 \ g}{10^6 \ \mu g} \times \frac{1 \ mol}{133.39 \ g} = 2.45 \times 10^{-8} \ mol \ C_2H_3Cl_3$$

    e.   molar mass of LiOH = 23.95 g

$$4.01 \ g \ LiOH \times \frac{1 \ mol}{23.95 \ g} = 0.167 \ mol \ LiOH$$

34.   a.   molar mass of $AlCl_3$ = 133.33 g

$$0.000314 \ g \ AlCl_3 \times \frac{1 \ mol}{133.33 \ g} = 2.36 \times 10^{-6} \ mol \ AlCl_3$$

    b.   molar mass of NaOH = 40.00 g

$$1.00 \ lb \times \frac{453.59 \ g}{1 \ lb} \times \frac{1 \ mol}{40.00 \ g} = 11.3 \ mol \ NaOH$$

    c.   molar mass of AsI_3 = 455.6 g

$$7.4 \ \mu g \ AsI_3 \times \frac{1 \ g}{10^6 \ \mu g} \times \frac{1 \ mol}{455.6 \ g} = 1.6 \times 10^{-8} \ mol \ AsI_3$$

    d.   molar mass of $NaHCO_3$ = 84.01 g

$$44.2 \ kg \ NaHCO_3 \times \frac{1000 \ g}{1 \ kg} \times \frac{1 \ mol}{84.01 \ g} = 526 \ mol \ NaHCO_3$$

    e.   molar mass of K_2HPO_4 = 174.2 g

$$6.21 \ g \ K_2HPO_4 \times \frac{1 \ mol}{174.2 \ g} = 0.0357 \ mol \ K_2HPO_4$$

36.   a.   molar mass of AlI_3 = 407.7 g

$$1.50 \ mol \ AlI_3 \times \frac{407.7 \ g}{1 \ mol} = 612 \ g \ AlI_3$$

    b.   molar mass of C_6H_6 = 78.11 g

$$1.91 \times 10^{-3} \ mol \ C_6H_6 \times \frac{78.11 \ g}{1 \ mol} = 0.149 \ g \ C_6H_6$$

    c.   molar mass of $C_6H_{12}O_6$ = 180.2 g

$$4.00 \ mol \ C_6H_{12}O_6 \times \frac{180.2 \ g}{1 \ mol} = 721 \ g \ C_6H_{12}O_6$$

Chapter 9   Chemical Composition   79

d. molar mass of C_2H_5OH = 46.07 g

$$4.56 \times 10^5 \text{ mol } C_2H_5OH \times \frac{46.07 \text{ g}}{1 \text{ mol}} = 2.10 \times 10^7 \text{ g } C_2H_5OH$$

e. molar mass of $Ca(NO_3)_2$ = 164.1 g

$$2.27 \text{ mol } Ca(NO_3)_2 \times \frac{164.1 \text{ g}}{1 \text{ mol}} = 373 \text{ g } Ca(NO_3)_2$$

38. a. molar mass of CO_2 = 44.01 g

$$1.27 \text{ mmol} \times \frac{1 \text{ mol}}{10^3 \text{ mmol}} \times \frac{44.01 \text{ g}}{1 \text{ mol}} = 0.0559 \text{ g } CO_2$$

b. molar mass of NCl_3 = 120.4 g

$$4.12 \times 10^3 \text{ mol } NCl_3 \times \frac{120.4 \text{ g}}{1 \text{ mol}} = 4.96 \times 10^5 \text{ g } NCl_3$$

c. molar mass of NH_4NO_3 = 80.05 g

$$0.00451 \text{ mol } NH_4NO_3 \times \frac{80.05 \text{ g}}{1 \text{ mol}} = 0.361 \text{ } NH_4NO_3$$

d. molar mass of H_2O = 18.02 g

$$18.0 \text{ mol } H_2O \times \frac{18.02 \text{ g}}{1 \text{ mol}} = 324 \text{ g } H_2O$$

e. molar mass of $CuSO_4$ = 159.6 g

$$62.7 \text{ mol } CuSO_4 \times \frac{159.6 \text{ g}}{1 \text{ mol}} = 1.00 \times 10^4 \text{ g } CuSO_4$$

40. a. $$4.29 \text{ mol } NO_2 \times \frac{6.022 \times 10^{23} \text{ molecules}}{1 \text{ mol}} = 2.58 \times 10^{24} \text{ molecules } NO_2$$

b. molar mass of NO_2 = 46.01 g

$$4.29 \text{ g} \times \frac{1 \text{ mol}}{46.01 \text{ g}} \times \frac{6.022 \times 10^{23} \text{ molec.}}{1 \text{ mol}} = 5.61 \times 10^{22} \text{ molec.}$$

c. $$1.95 \times 10^{-10} \text{ mol} \times \frac{6.022 \times 10^{23} \text{ molec.}}{1 \text{ mol}} = 1.17 \times 10^{14} \text{ molecules}$$

d. molar mass of HF = 20.008 g

$$1.95 \times 10^{-10} \text{ g} \times \frac{1 \text{ mol}}{20.008 \text{ g}} \times \frac{6.022 \times 10^{23} \text{ molec.}}{1 \text{ mol}} = 5.87 \times 10^{12} \text{ molec}$$

e. molar mass of NH_3 = 17.03 g

80 Chapter 9   Chemical Composition

$$4.61 \text{ g} \times \frac{1 \text{ mol}}{17.03 \text{ g}} \times \frac{6.022 \times 10^{23} \text{ molec.}}{1 \text{ mol}} = 1.63 \times 10^{23} \text{ molecules}$$

42. a. molar mass of Na_2SO_4 = 142.1 g

$$2.01 \text{ g } Na_2SO_4 \times \frac{1 \text{ mol } Na_2SO_4}{142.1 \text{ g}} \times \frac{1 \text{ mol S}}{1 \text{ mol } Na_2SO_4} = 0.0141 \text{ mol S}$$

b. molar mass of Na_2SO_3 = 126.1 g

$$2.01 \text{ g } Na_2SO_3 \times \frac{1 \text{ mol } Na_2SO_3}{126.1 \text{ g}} \times \frac{1 \text{ mol S}}{1 \text{ mol } Na_2SO_3} = 0.0159 \text{ mol S}$$

c. molar mass of Na_2S = 78.05 g

$$2.01 \text{ g } Na_2S \times \frac{1 \text{ mol } Na_2S}{78.05 \text{ g}} \times \frac{1 \text{ mol S}}{1 \text{ mol } Na_2S} = 0.0258 \text{ mol S}$$

d. molar mass of $Na_2S_2O_3$ = 158.1 g

$$2.01 \text{ g } Na_2S_2O_3 \times \frac{1 \text{ mol } Na_2S_2O_3}{158.1 \text{ g}} \times \frac{2 \text{ mol S}}{1 \text{ mol } Na_2S_2O_3} = 0.0254 \text{ mol S}$$

44. less than

46. a. mass of Na present =   2(22.99 g) =   45.98 g
       mass of S present =   32.07 g =   32.07 g
       mass of O present =   4(16.00 g) =   64.00 g
       molar mass of Na_2SO_4 =   142.05 g

$$\%Na = \frac{45.98 \text{ g Na}}{142.05 \text{ g}} \times 100 = 32.37\% \text{ Na}$$

$$\%S = \frac{32.07 \text{ g S}}{142.05 \text{ g}} \times 100 = 22.57\% \text{ S}$$

$$\%O = \frac{64.00 \text{ g O}}{142.05 \text{ g}} \times 100 = 45.05\% \text{ O}$$

b. mass of Na present =   2(22.99 g) =   45.98 g
    mass of S present =   32.07 g =   32.07 g
    mass of O present =   3(16.00 g) =   48.00 g
    molar mass of Na_2SO_3 =   126.05 g

$$\%Na = \frac{45.98 \text{ g Na}}{126.05 \text{ g}} \times 100 = 36.46\% \text{ Na}$$

$$\%S = \frac{32.07 \text{ g S}}{126.05 \text{ g}} \times 100 = 25.43\% \text{ S}$$

$$\%O = \frac{48.00 \text{ g O}}{126.05 \text{ g}} \times 100 = 38.07\% \text{ O}$$

c. mass of Na present = 2(22.99 g) = 45.98 g
   mass of S present = 32.07 g = 32.07 g
   ──
   molar mass of Na₂S = 78.05 g

$$\%Na = \frac{45.98 \text{ g Na}}{78.05 \text{ g}} \times 100 = 58.91\% \text{ Na}$$

$$\%S = \frac{32.07 \text{ g S}}{78.05 \text{ g}} \times 100 = 41.09\% \text{ S}$$

d. mass of Na present = 2(22.99 g) = 45.98 g
   mass of S present = 2(32.07 g) = 64.14 g
   mass of O present = 3(16.00 g) = 48.00 g
   ──
   molar mass of Na₂S₂O₃ = 158.12 g

$$\%Na = \frac{45.98 \text{ g Na}}{158.12 \text{ g}} \times 100 = 29.08\% \text{ Na}$$

$$\%S = \frac{64.14 \text{ g S}}{158.12 \text{ g}} \times 100 = 40.56\% \text{ S}$$

$$\%O = \frac{48.00 \text{ g O}}{158.12 \text{ g}} \times 100 = 30.36\% \text{ O}$$

e. mass of K present = 3(39.10 g) = 117.3 g
   mass of P present = 30.97 g = 30.97 g
   mass of O present = 4(16.00 g) = 64.00 g
   ──
   molar mass of K₃PO₄ = 212.3 g

$$\%K = \frac{117.3 \text{ g K}}{212.3 \text{ g}} \times 100 = 55.25\% \text{ K}$$

$$\%P = \frac{30.97 \text{ g P}}{212.3 \text{ g}} \times 100 = 14.59\% \text{ P}$$

$$\%O = \frac{64.00 \text{ g O}}{212.3 \text{ g}} \times 100 = 30.15\% \text{ O}$$

f. mass of K present = 2(39.10 g) = 78.20 g
   mass of H present = 1.008 g = 1.008 g
   mass of P present = 30.97 g = 30.97 g
   mass of O present = 4(16.00 g) = 64.00 g
   ──
   molar mass of K₂HPO₄ = 174.178 g = 174.18 g

82 Chapter 9   Chemical Composition

$$\%K = \frac{78.20 \text{ g K}}{174.18 \text{ g}} \times 100 = 44.90\% \text{ K}$$

$$\%H = \frac{1.008 \text{ g H}}{174.18 \text{ g}} \times 100 = 0.5787\% \text{ H}$$

$$\%P = \frac{30.97 \text{ g P}}{174.18 \text{ g}} \times 100 = 17.78\% \text{ P}$$

$$\%O = \frac{64.00 \text{ g O}}{174.18 \text{ g}} \times 100 = 36.74\% \text{ O}$$

g. mass of K present =   39.10 g =   39.10 g
    mass of H present =   2(1.008 g) =   2.016 g
    mass of P present =   30.97 g =   30.97 g
    mass of O present =   4(16.00 g) = 64.00 g
    molar mass of KH_2PO_4 =   136.09 g

$$\%K = \frac{39.10 \text{ g K}}{136.09 \text{ g}} \times 100 = 28.73\% \text{ K}$$

$$\%H = \frac{2.016 \text{ g H}}{136.09 \text{ g}} \times 100 = 1.481\% \text{ H}$$

$$\%P = \frac{30.97 \text{ g P}}{136.09 \text{ g}} \times 100 = 22.76\% \text{ P}$$

$$\%O = \frac{64.00 \text{ g O}}{136.09 \text{ g}} \times 100 = 47.03\% \text{ O}$$

h. mass of K present =   3(39.10 g) = 117.3 g
    mass of P present =   30.97 g =   30.97 g
    molar mass of K_3P =   148.27 g

$$\%K = \frac{117.3 \text{ g K}}{148.27 \text{ g}} \times 100 = 79.11\% \text{ K}$$

$$\%P = \frac{30.97 \text{ g P}}{148.3 \text{ g}} \times 100 = 20.89\% \text{ P}$$

48. a. molar mass of Ca_3N_2 = 148.26 g

$$\%Ca = \frac{120.24 \text{ g Ca}}{148.26 \text{ g}} \times 100 = 81.10\% \text{ Ca}$$

b. molar mass of $KMnO_4$ = 158.04 g

$$\%K = \frac{39.10 \text{ g K}}{158.04 \text{ g}} \times 100 = 24.74\% \text{ K}$$

c. molar mass of $SnBr_2$ = 278.5 g

$$\%Sn = \frac{118.7 \text{ g Sn}}{278.5 \text{ g}} \times 100 = 42.62\% \text{ Sn}$$

d. molar mass of Al_2S_3 = 150.17 g

$$\%Al = \frac{53.96 \text{ g Al}}{150.17 \text{ g}} \times 100 = 35.93\% \text{ Al}$$

e. molar mass C_2H_2 = 26.04 g

$$\%C = \frac{24.02 \text{ g C}}{26.04 \text{ g}} \times 100 = 92.24\% \text{ C}$$

f. molar mass of $(NH_4)_3P$ = 85.10 g

$$\%N = \frac{42.03 \text{ g N}}{85.10 \text{ g}} \times 100 = 49.39\% \text{ N}$$

g. molar mass of FeI_3 = 436.6 g

$$\%Fe = \frac{55.85 \text{ g Fe}}{436.6 \text{ g}} \times 100 = 12.79\% \text{ Fe}$$

h. molar mass of P_4O_{10} = 283.9 g

$$\%P = \frac{123.9 \text{ g P}}{283.9 \text{ g}} \times 100 = 43.64\% \text{ P}$$

50. a. molar mass of $FeCl_3$ = 162.2 g

$$\%Fe = \frac{55.85 \text{ g Fe}}{162.2 \text{ g}} \times 100 = 34.43\% \text{ Fe}$$

b. molar mass of OF_2 = 54.00 g

$$\%O = \frac{16.00 \text{ g O}}{54.00 \text{ g}} \times 100 = 29.63\% \text{ O}$$

c. molar mass of C_6H_6 = 78.11 g

84   Chapter 9   Chemical Composition

$$\%C = \frac{72.06 \text{ g C}}{78.11 \text{ g}} \times 100 = 92.25\% \text{ C}$$

d. molar mass of NH_4ClO_4 = 117.5 g

$$\%N = \frac{14.01 \text{ g N}}{117.5 \text{ g}} \times 100 = 11.92\% \text{ N}$$

e. molar mass of Ag_2O = 231.8 g

$$\%Ag = \frac{215.8 \text{ g Ag}}{231.8 \text{ g}} \times 100 = 93.10\% \text{ Ag}$$

f. molar mass of $CoCl_2$ = 129.83 g

$$\%Co = \frac{58.93 \text{ g Co}}{129.83 \text{ g}} \times 100 = 45.39\% \text{ Co}$$

g. molar mass of N_2O_4 = 92.02 g

$$\%N = \frac{28.02 \text{ g N}}{92.02 \text{ g}} \times 100 = 30.45\% \text{ N}$$

h. molar mass of $MnCl_2$ = 125.8 g

$$\%Mn = \frac{54.94 \text{ g Mn}}{125.8 \text{ g}} \times 100 = 43.66\% \text{ Mn}$$

52. a. molar mass of $Ca_3(PO_4)_2$ = 310.18 g
    molar mass of PO_4^{3-} ion = 94.97 g

$$\% PO_4^{3-} = \frac{189.9 \text{ g } PO_4^{3-}}{310.18 \text{ g}} \times 100 = 61.22\% \text{ } PO_4^{3-}$$

b. molar mass of $CdSO_4$ = 208.5 g
   molar mass of SO_4^{2-} ion = 96.07 g

$$\% SO_4^{2-} = \frac{96.07 \text{ g } SO_4^{2-}}{208.5 \text{ g}} \times 100 = 46.08\% \text{ } SO_4^{2-}$$

c. molar mass of $Fe_2(SO_4)_3$ = 399.9 g
   molar mass of SO_4^{2-} ion = 96.07 g

$$\% SO_4^{2-} = \frac{288.2 \text{ g } SO_4^{2-}}{399.9 \text{ g}} \times 100 = 72.07\% \text{ } SO_4^{2-}$$

d. molar mass of $MnCl_2$ = 125.84 g
   molar mass of Cl^- ion = 35.45 g

$$\% Cl^- = \frac{70.90 \text{ g } Cl^-}{125.84 \text{ g}} \times 100 = 56.34\% \text{ } Cl^-$$

54. The empirical formula represents the smallest whole number ratio of the elements present in a compound. The molecular formula indicates the actual number of atoms of each element found in a molecule of the substance.

56. a. yes (each of these has empirical formula CH)

　　b. no (the number of hydrogen atoms is wrong)

　　c. yes (both have empirical formula NO_2)

　　d. no (the number of hydrogen atoms is wrong)

58. $1.00 \text{ g Ba} \times \dfrac{1 \text{ mol}}{137.3 \text{ g Ba}} = 0.007283 \text{ mol Ba}$

If the oxide sample weighs 1.117 g, then (1.117 − 1.000) = 0.117 g O reacted.

$0.117 \text{ g O} \times \dfrac{1 \text{ mol}}{16.00 \text{ g O}} = 0.00730 \text{ mol O}$

Since we have 0.0073 mol of both barium and oxygen present (i.e., the *same* number of moles), the empirical formula must be BaO.

60. Consider having 100.0 g of the compound. Then the percentages of the elements present are numerically equal to their masses in grams.

$58.84 \text{ g Ba} \times \dfrac{1 \text{ mol}}{137.3 \text{ g Ba}} = 0.4286 \text{ mol Ba}$

$13.74 \text{ g S} \times \dfrac{1 \text{ mol}}{32.07 \text{ g S}} = 0.4284 \text{ mol S}$

$27.43 \text{ g O} \times \dfrac{1 \text{ mol}}{16.00 \text{ g O}} = 1.714 \text{ mol O}$

Dividing each number of moles by the smallest number of moles (0.4284 mol S) gives

$\dfrac{0.4285 \text{ mol Ba}}{0.4284} = 1.000 \text{ mol Ba}$

$\dfrac{0.4284 \text{ mol S}}{0.4284} = 0.9998 \text{ mol S}$

$\dfrac{1.714 \text{ mol O}}{0.4284} = 4.000 \text{ mol O}$

The empirical formula is $BaSO_4$.

86 Chapter 9 Chemical Composition

62. Consider 100.0 g of the compound.

$$29.16 \text{ g N} \times \frac{1 \text{ mol}}{14.01 \text{ g N}} = 2.081 \text{ mol N}$$

$$8.392 \text{ g H} \times \frac{1 \text{ mol}}{1.008 \text{ g H}} = 8.325 \text{ mol H}$$

$$12.50 \text{ g C} \times \frac{1 \text{ mol}}{12.01 \text{ g C}} = 1.041 \text{ mol C}$$

$$49.95 \text{ g O} \times \frac{1 \text{ mol}}{16.00 \text{ g O}} = 3.122 \text{ mol O}$$

Dividing each number of moles by the smallest (1.048 mol C) gives

$$\frac{2.081 \text{ mol N}}{1.041} = 1.999 \text{ mol N}$$

$$\frac{8.325 \text{ mol H}}{1.041} = 7.997 \text{ mol H}$$

$$\frac{1.041 \text{ mol C}}{1.041} = 1.000 \text{ mol C}$$

$$\frac{3.121 \text{ mol O}}{1.041} = 2.998 \text{ mol O}$$

The empirical formula is $N_2H_8CO_3$ [i.e, $(NH_4)_2CO_3$].

64. Consider 100.0 g of the compound.

$$55.06 \text{ g Co} \times \frac{1 \text{ mol}}{58.93 \text{ g Co}} = 0.9343 \text{ mol Co}$$

If the sulfide of cobalt is 55.06% Co, then it is 44.94% S by mass.

$$44.94 \text{ g S} \times \frac{1 \text{ mol}}{32.07 \text{ g S}} = 1.401 \text{ mol S}$$

Dividing each number of moles by the smaller (0.9343 mol Co) gives

$$\frac{0.9343 \text{ mol Co}}{0.9343} = 1.000 \text{ mol Co}$$

$$\frac{1.401 \text{ mol S}}{0.9343} = 1.500 \text{ mol S}$$

Multiplying by two, to convert to whole numbers of moles, gives the empirical formula for the compound as Co_2S_3.

66. $10.00 \text{ g Cu} \times \dfrac{1 \text{ mol}}{63.55 \text{ g Cu}} = 0.1574 \text{ mol Cu}$

$2.52 \text{ g O} \times \dfrac{1 \text{ mol}}{16.00 \text{ g O}} = 0.1575 \text{ mol O}$

Since the numbers of moles are equal, the empirical formula is just CuO.

68. Consider 100.0 g of the compound.

$32.13 \text{ g Al} \times \dfrac{1 \text{ mol}}{26.98 \text{ g Al}} = 1.191 \text{ mol Al}$

$67.87 \text{ g F} \times \dfrac{1 \text{ mol}}{19.00 \text{ g F}} = 3.572 \text{ mol F}$

Dividing each number of moles by the smaller number (1.191 mol Al) gives

$\dfrac{1.191 \text{ mol Al}}{1.191} = 1.000 \text{ mol Al}$

$\dfrac{3.572 \text{ mol F}}{1.191} = 2.999 \text{ mol F}$

The empirical formula is AlF_3

70. Consider 100.0 g of the compound.

$59.78 \text{ g Li} \times \dfrac{1 \text{ mol Li}}{6.941 \text{ g Li}} = 8.613 \text{ mol Li}$

$40.22 \text{ g N} \times \dfrac{1 \text{ mol N}}{14.01 \text{ g N}} = 2.871 \text{ mol N}$

Dividing each number of moles by the smaller number of moles (2.871 mol N) gives

$\dfrac{8.613 \text{ mol Li}}{2.871} = 3.000 \text{ mol Li}$

$\dfrac{2.871 \text{ mol N}}{2.871} = 1.000 \text{ mol N}$

The empirical formula is Li_3N.

72. Consider 100.0 g of the compound.

$15.77 \text{ g Al} \times \dfrac{1 \text{ mol Al}}{26.98 \text{ g Al}} = 0.5845 \text{ mol Al}$

$28.11 \text{ g S} \times \dfrac{1 \text{ mol S}}{32.07 \text{ g S}} = 0.8765 \text{ mol S}$

88   Chapter 9   Chemical Composition

$$56.12 \text{ g O} \times \frac{1 \text{ mol O}}{16.00 \text{ g O}} = 3.508 \text{ mol O}$$

Dividing each number of moles by the smallest number of moles (0.5845 mol Al) gives

$$\frac{0.5845 \text{ mol Al}}{0.5845} = 1.000 \text{ mol Al}$$

$$\frac{0.8765 \text{ mol S}}{0.5845} = 1.500 \text{ mol S}$$

$$\frac{3.508 \text{ mol O}}{0.5845} = 6.002 \text{ mol O}$$

Multiplying these relative numbers of moles by 2 to give whole numbers gives the empirical formula as $Al_2S_3O_{12}$.

74. Consider 100.0 g of the compound.

$$21.60 \text{ g Na} \times \frac{1 \text{ mol Na}}{22.99 \text{ g Na}} = 0.9395 \text{ mol Na}$$

$$33.31 \text{ g Cl} \times \frac{1 \text{ mol Cl}}{35.45 \text{ g Cl}} = 0.9396 \text{ mol Cl}$$

$$45.09 \text{ g O} \times \frac{1 \text{ mol O}}{16.00 \text{ g O}} = 2.818 \text{ mol O}$$

Dividing each number of moles by the smallest number of moles (0.9395 mol Na) gives

$$\frac{0.9395 \text{ mol Na}}{0.9395} = 1.000 \text{ mol Na}$$

$$\frac{0.9396 \text{ mol Cl}}{0.9395} = 1.000 \text{ mol Cl}$$

$$\frac{2.818 \text{ mol O}}{0.9395} = 2.999 \text{ mol O}$$

The empirical formula is $NaClO_3$.

76. If only the empirical formula is known, the molar mass of the substance must be determined before the molecular formula can be calculated.

molecular formula is $(CH_2O)_3 = C_3H_6O_3$.

78. empirical formula mass of CH = 13 g

$$n = \frac{\text{molar mass}}{\text{empirical formula mass}} = \frac{78 \text{ g}}{13 \text{ g}} = 6$$

The molecular formula is $(CH)_6$ or C_6H_6.

80. empirical formula mass of NO_2 = 46 g

$$n = \frac{\text{molar mass}}{\text{empirical formula mass}} = \frac{92 \text{ g}}{46 \text{ g}} = 2$$

molecular formula is $(NO_2)_2 = N_2O_4$.

82. Consider 100.0 g of the compound.

$$65.45 \text{ g C} \times \frac{1 \text{ mol C}}{12.01 \text{ g C}} = 5.450 \text{ mol C}$$

$$5.492 \text{ g H} \times \frac{1 \text{ mol H}}{1.008 \text{ g H}} = 5.448 \text{ mol H}$$

$$29.06 \text{ g O} \times \frac{1 \text{ mol O}}{16.00 \text{ g O}} = 1.816 \text{ mol O}$$

Dividing each number of moles by the smallest number of moles (1.816 mol O) gives

$$\frac{5.450 \text{ mol C}}{1.816} = 3.001 \text{ mol C}$$

$$\frac{5.448 \text{ mol H}}{1.816} = 3.000 \text{ mol H}$$

$$\frac{1.816 \text{ mol O}}{1.816} = 1.000 \text{ mol O}$$

The empirical formula is C_3H_3O, and the empirical formula mass is approximately 55 g.

$$n = \frac{\text{molar mass}}{\text{empirical formula mass}} = \frac{110 \text{ g}}{55 \text{ g}} = 2$$

The molecular formula is $(C_3H_3O)_2 = C_6H_6O_2$.

84.
5.00 g Al	0.185 mol	1.12×10^{23} atoms
0.140 g Fe	0.00250 mol	1.51×10^{21} atoms
2.7×10^2 g Cu	4.3 mol	2.6×10^{24} atoms
0.00250 g Mg	1.03×10^{-4} mol	6.19×10^{19} atoms
0.062 g Na	2.7×10^{-3} mol	1.6×10^{21} atoms
3.95×10^{-18} g U	1.66×10^{-20} mol	1.00×10^4 atoms

90   Chapter 9   Chemical Composition

86. mass of 2 mol X =   2(41.2 g) =   82.4 g
     mass of 1 mol Y =   57.7 g =   57.7 g
     mass of 3 mol Z =   3(63.9 g) =   191.7 g
     molar mass of X_2YZ_3 =   331.8 g

$$\% \text{ X} = \frac{82.4 \text{ g}}{331.8 \text{ g}} \times 100 = 24.8\% \text{ X}$$

$$\% \text{ Y} = \frac{57.7 \text{ g}}{331.8 \text{ g}} \times 100 = 17.4\% \text{ Y}$$

$$\% \text{ Z} = \frac{191.7 \text{ g}}{331.8 \text{ g}} \times 100 = 57.8\% \text{ Z}$$

If the molecular formula were actually $X_4Y_2Z_6$, the percentage composition would be the same, and the *relative* mass of each element present would not change. The molecular formula is always a whole number multiple of the empirical formula.

88. For the first compound (*restricted* amount of oxygen)

$$2.118 \text{ g Cu} \times \frac{1 \text{ mol Cu}}{63.54 \text{ g Cu}} = 0.03333 \text{ mol Cu}$$

$$0.2666 \text{ g O} \times \frac{1 \text{ mol O}}{16.00 \text{ g O}} = 0.01666 \text{ mol O}$$

Since the number of moles of Cu (0.03333 mol) is twice the number of moles of O (0.01666 mol), the empirical formula is Cu_2O.

For the second compound (stream of pure oxygen)

$$2.118 \text{ g Cu} \times \frac{1 \text{ mol Cu}}{63.54 \text{ g Cu}} = 0.03333 \text{ mol Cu}$$

$$0.5332 \text{ g O} \times \frac{1 \text{ mol O}}{16.00 \text{ g O}} = 0.03333 \text{ mol O}$$

Since the numbers of moles are the same, the empirical formula is CuO.

90. a.   molar mass of $Pb(NO_3)_2$ = 331.2 g

$$2.37 \text{ g} \times \frac{1 \text{ mol}}{331.2 \text{ g}} = 7.15 \times 10^{-3} \text{ mol Pb(NO}_3)_2$$

$$7.15 \times 10^{-3} \text{ mol Pb(NO}_3)_2 \times \frac{1 \text{ mol Pb}}{1 \text{ mol Pb(NO}_3)_2} = 7.15 \times 10^{-3} \text{ mol Pb}$$

$$7.15 \times 10^{-3} \text{ mol Pb} \times \frac{6.022 \times 10^{23} \text{ atoms}}{1 \text{ mol}} = 4.31 \times 10^{21} \text{ Pb atoms}$$

$$7.15 \times 10^{-3} \text{ mol Pb(NO}_3)_2 \times \frac{2 \text{ mol N}}{1 \text{ mol Pb(NO}_3)_2} = 0.0143 \text{ mol N}$$

$$0.0143 \text{ mol N} \times \frac{6.022 \times 10^{23} \text{ atoms}}{1 \text{ mol}} = 8.61 \times 10^{21} \text{ N atoms}$$

$$7.15 \times 10^{-3} \text{ mol Pb(NO}_3)_2 \times \frac{6 \text{ mol O}}{1 \text{ mol Pb(NO}_3)_2} = 0.0429 \text{ mol O}$$

$$0.0429 \text{ mol O} \times \frac{6.022 \times 10^{23} \text{ atoms}}{1 \text{ mol}} = 2.59 \times 10^{22} \text{ O atoms}$$

b. molar mass O_2 = 32.00 g

$$22.4 \text{ g O}_2 \times \frac{1 \text{ mol}}{32.00 \text{ g}} = 0.700 \text{ mol O}_2$$

$$0.700 \text{ mol O}_2 \times \frac{2 \text{ mol O atoms}}{1 \text{ mol O}_2} = 1.40 \text{ mol O atoms}$$

$$1.40 \text{ mol O} \times \frac{6.022 \times 10^{23} \text{ atoms}}{1 \text{ mol}} = 8.43 \times 10^{23} \text{ O atoms}$$

c. molar mass S_8 = 256.56 g

101.9 mg = 0.1019 g

$$0.1019 \text{ g} \times \frac{1 \text{ mol S}_8}{256.56 \text{ g}} = 3.972 \times 10^{-4} \text{ mol S}_8$$

$$3.972 \times 10^{-4} \text{ mol S}_8 \times \frac{8 \text{ mol S}}{1 \text{ mol S}_8} = 3.177 \times 10^{-3} \text{ mol S}$$

$$3.177 \times 10^{-3} \text{ mol S} \times \frac{6.022 \times 10^{23} \text{ S atoms}}{1 \text{ mol}} = 1.913 \times 10^{21} \text{ S atoms}$$

d. molar mass UF_6 = 352.0 g

43.7 μg = 43.7 × 10^{-6} g

$$43.7 \times 10^{-6} \text{ g} \times \frac{1 \text{ mol}}{352.0 \text{ g}} = 1.24 \times 10^{-7} \text{ mol UF}_6$$

$$1.24 \times 10^{-7} \text{ mol UF}_6 \times \frac{1 \text{ mol U}}{1 \text{ mol UF}_6} = 1.24 \times 10^{-7} \text{ mol U}$$

$$1.24 \times 10^{-7} \text{ mol U} \times \frac{6.022 \times 10^{23} \text{ U atoms}}{1 \text{ mol}} = 7.47 \times 10^{16} \text{ U atoms}$$

$$1.24 \times 10^{-7} \text{ mol UF}_6 \times \frac{6 \text{ mol F}}{1 \text{ mol UF}_6} = 7.44 \times 10^{-7} \text{ mol F}$$

$$7.44 \times 10^{-7} \text{ mol F} \times \frac{6.022 \times 10^{23} \text{ F atoms}}{1 \text{ mol}} = 4.48 \times 10^{17} \text{ F atoms}$$

92. a. molar mass of C_3O_2 = 3(12.01 g) + 2(16.00 g) = 68.03 g

$$\% \text{ C} = \frac{36.03 \text{ g}}{68.03 \text{ g}} = 52.96\% \text{ C}$$

$$7.819 \text{ g } C_3O_2 \times \frac{52.96 \text{ g C}}{100.0 \text{ g } C_3O_2} = 4.141 \text{ g C}$$

$$4.141 \text{ g C} \times \frac{6.022 \times 10^{23} \text{ C atoms}}{12.01 \text{ g C}} = 2.076 \times 10^{23} \text{ C atoms}$$

b. molar mass of CO = 12.01 g + 16.00 g = 28.01 g

$$\% \text{ C} = \frac{12.01 \text{ g}}{28.01 \text{ g}} \times 100 = 42.9\% \text{ C}$$

$$1.53 \times 10^{21} \text{ molecules CO} \times \frac{1 \text{ C atom}}{1 \text{ molecule CO}} = 1.53 \times 10^{21} \text{ C atoms}$$

$$1.53 \times 10^{21} \text{ C atoms} \times \frac{12.01 \text{ g C}}{6.022 \times 10^{23} \text{ C atoms}} = 0.0305 \text{ g C}$$

c. molar mass of C_6H_6O = 6(12.01 g) + 6(1.008 g) + 16.00 g = 94.11 g

$$\% \text{ C} = \frac{72.06 \text{ g}}{94.11 \text{ g}} \times 100 = 76.57\% \text{ C}$$

$$0.200 \text{ mol } C_6H_6O \times \frac{6 \text{ mol C}}{1 \text{ mol } C_6H_6O} = 1.20 \text{ mol C}$$

$$1.20 \text{ mol C} \times \frac{12.01 \text{ g C}}{1 \text{ mol C}} = 14.4 \text{ g C}$$

$$14.4 \text{ g C} \times \frac{6.022 \times 10^{23} \text{ C atoms}}{12.01 \text{ g C}} = 7.22 \times 10^{23} \text{ C atoms}$$

94. $1.00 \text{ g Ne} \times \dfrac{207.2 \text{ g Pb}}{20.18 \text{ g Ne}} = 10.3 \text{ g Pb}$

96. $5.00 \text{ g Te} \times \dfrac{200.6 \text{ g Hg}}{127.6 \text{ g Te}} = 7.86 \text{ g Hg}$

98. 153.8 g CCl$_4$ = 6.022 × 10^{23} molecules CCl$_4$

$$1 \text{ molecule} \times \frac{153.8 \text{ g}}{6.022 \times 10^{23} \text{ molecules}} = 2.554 \times 10^{-22} \text{ g}$$

100. a. molar mass of C$_2$H$_5$O$_2$N = 75.07 g

   $$\text{mass fraction N} = \frac{14.01 \text{ g}}{75.07 \text{ g}}$$

   $$5.000 \text{ g} \times \frac{14.01 \text{ g}}{75.07 \text{ g}} = 0.9331 \text{ g N}$$

 b. molar mass of Mg$_3$N$_2$ = 3(24.31 g) + 2(14.01 g) = 100.95 g

   $$\text{mass fraction N} = \frac{28.02 \text{ g}}{100.95 \text{ g}}$$

   $$5.000 \text{ g} \times \frac{28.02 \text{ g}}{100.95 \text{ g}} = 1.388 \text{ g N}$$

 c. molar mass of Ca(NO$_3$)$_2$ =
   40.08 g + 2(14.01 g) + 6(16.00 g) = 164.10 g

   $$\text{mass fraction N} = \frac{28.02 \text{ g}}{164.10 \text{ g}}$$

   $$5.000 \text{ g} \times \frac{28.02 \text{ g}}{164.10 \text{ g}} = 0.8537 \text{ g N}$$

 d. molar mass of N$_2$O$_4$ = 2(14.01 g) + 4(16.00 g) = 92.02 g

   $$\text{mass fraction N} = \frac{28.02 \text{ g}}{92.02 \text{ g}}$$

   $$5.000 \text{ g} \times \frac{28.02 \text{ g}}{92.02 \text{ g}} = 1.522 \text{ g N}$$

102. Consider 100.0 g of the compound.

   $$16.39 \text{ g Mg} \times \frac{1 \text{ mol Mg}}{24.31 \text{ g Mg}} = 0.6742 \text{ mol Mg}$$

   $$18.89 \text{ g N} \times \frac{1 \text{ mol N}}{14.01 \text{ g N}} = 1.348 \text{ mol N}$$

   $$64.72 \text{ g O} \times \frac{1 \text{ mol O}}{16.00 \text{ g O}} = 4.045 \text{ mol O}$$

   Dividing each number of moles by the smallest number of moles

94   Chapter 9   Chemical Composition

$$\frac{0.6742 \text{ mol Mg}}{0.6742} = 1.000 \text{ mol Mg}$$

$$\frac{1.348 \text{ mol N}}{0.6742} = 1.999 \text{ mol N}$$

$$\frac{4.045 \text{ mol O}}{0.6742} = 5.999 \text{ mol O}$$

The empirical formula is MgN_2O_6.

104. (weighted) average

106. 1.98×10^{13} amu $\times \dfrac{1 \text{ Na atom}}{22.99 \text{ amu}} = 8.61 \times 10^{11}$ Na atoms

3.01×10^{23} Na atoms $\times \dfrac{22.99 \text{ amu}}{1 \text{ Na atom}} = 6.92 \times 10^{24}$ amu

108. a.   5.0 mol K $\times \dfrac{39.10 \text{ g}}{1 \text{ mol}} = 195 \text{ g} = 2.0 \times 10^2$ g K

b.   0.000305 mol Hg $\times \dfrac{200.6 \text{ g}}{1 \text{ mol}} = 0.0612$ g Hg

c. 2.31×10^{-5} mol Mn $\times \dfrac{54.94 \text{ g}}{1 \text{ mol}} = 1.27 \times 10^{-3}$ g Mn

d.   10.5 mol P $\times \dfrac{30.97 \text{ g}}{1 \text{ mol}} = 325$ g P

e. 4.9×10^4 mol Fe $\times \dfrac{55.85 \text{ g}}{1 \text{ mol}} = 2.7 \times 10^6$ g Fe

f.   125 mol Li $\times \dfrac{6.941 \text{ g}}{1 \text{ mol}} = 868$ g Li

g.   0.01205 mol F $\times \dfrac{19.00 \text{ g}}{1 \text{ mol}} = 0.2290$ g F

110. a.   mass of 3 mol C =   3(12.01 g) =   36.03
          mass of 8 mol H =   8(1.008 g) =   8.064 g
          molar mass of C_3H_8 =   44.09 g

b.   mass of 2 mol Fe =   2(55.85 g) = 111.7 g
     mass of 3 mol S =   3(32.07 g) = 96.21 g
     mass of 12 mol O = 12(16.00 g) = 192.0 g
     molar mass of $Fe_2(SO_4)_3$ =   399.9 g

c. $\quad$ mass of 2 mol N = $\quad$ 2(14.01 g) = $\quad$ 28.02 g
$\quad\quad$ mass of 5 mol O = $\quad$ 5(16.00 g) = $\quad$ 80.00 g
$\quad\quad$ molar mass of N_2O_5 = $\quad\quad\quad\quad\quad\quad$ 108.02 g = 108.0 g

d. $\quad$ mass of 12 mol C = 12(12.01 g) = 144.1 g
$\quad\quad$ mass of 22 mol H = 22(1.008 g) = 22.18 g
$\quad\quad$ mass of 11 mol O = 11(16.00 g) = 176.0 g
$\quad\quad$ molar mass of $C_{12}H_{22}O_{11}$ = $\quad\quad\quad\quad$ 342.3 g

e. $\quad$ mass of 2 mol N = $\quad$ 2(14.01 g) = $\quad$ 28.02 g
$\quad\quad$ mass of 8 mol H = $\quad$ 8(1.008 g) = $\quad$ 8.064 g
$\quad\quad$ mass of 1 mol C = $\quad$ 12.01 g = $\quad$ 12.01 g
$\quad\quad$ mass of 3 mol O = $\quad$ 3(16.00 g) = $\quad$ 48.00 g
$\quad\quad$ molar mass of $(NH_4)_2CO_3$ = $\quad\quad\quad\quad$ 96.09 g

f. $\quad$ mass of 1 mol Ba = $\quad$ 137.3 g $\quad$ = $\quad$ 137.3 g
$\quad\quad$ mass of 2 mol Cl = $\quad$ 2(35.45 g) = $\quad$ 70.90 g
$\quad\quad$ molar mass of $BaCl_2$ = $\quad\quad\quad\quad\quad$ 208.2 g

112. a. $\quad$ molar mass of $(NH_4)_2S$ = 68.15 g

$$21.2 \text{ g} \times \frac{1 \text{ mol}}{68.15 \text{ g}} = 0.311 \text{ mol } (NH_4)_2S$$

b. $\quad$ molar mass of $Ca(NO_3)_2$ = 164.1 g

$$44.3 \text{ g} \times \frac{1 \text{ mol}}{164.1 \text{ g}} = 0.270 \text{ mol } Ca(NO_3)_2$$

c. $\quad$ molar mass of Cl_2O = 86.9 g

$$4.35 \text{ g} \times \frac{1 \text{ mol}}{86.9 \text{ g}} = 0.0501 \text{ mol } Cl_2O$$

d. $\quad$ 1.0 lb = 454 g
$\quad\quad$ molar mass of $FeCl_3$ = 162.2

$$454 \text{ g} \times \frac{1 \text{ mol}}{162.2 \text{ g}} = 2.8 \text{ mol } FeCl_3$$

e. $\quad$ 1.0 kg = 1.0 × 10³ g
$\quad\quad$ molar mass of $FeCl_3$ = 162.2 g

$$1.0 \times 10^3 \text{ g} \times \frac{1 \text{ mol}}{162.2 \text{ g}} = 6.2 \text{ mol } FeCl_3$$

114.
a. molar mass of $CuSO_4$ = 159.62 g

$$2.6 \times 10^{-2} \text{ mol} \times \frac{159.62 \text{ g}}{1 \text{ mol}} = 4.2 \text{ g } CuSO_4$$

b. molar mass of C_2F_4 = 100.0 g

$$3.05 \times 10^3 \text{ mol} \times \frac{100.0 \text{ g}}{1 \text{ mol}} = 3.05 \times 10^5 \text{ g } C_2F_4$$

c. 7.83 millimol = 0.00783 mol

molar mass of C_5H_8 = 68.1 g

$$0.00783 \text{ mol} \times \frac{68.1 \text{ g}}{1 \text{ mol}} = 0.533 \text{ g } C_5H_8$$

d. molar mass of $BiCl_3$ = 315.3 g

$$6.30 \text{ mol} \times \frac{315.3 \text{ g}}{1 \text{ mol}} = 1.99 \times 10^3 \text{ g } BiCl_3$$

e. molar mass of $C_{12}H_{22}O_{11}$ = 342.3 g

$$12.2 \text{ mol} \times \frac{342.3 \text{ g}}{1 \text{ mol}} = 4.18 \times 10^3 \text{ g } C_{12}H_{22}O_{11}$$

116.
a. molar mass of $C_6H_{12}O_6$ = 180.2 g

$$3.45 \text{ g} \times \frac{6.022 \times 10^{23} \text{ molecules}}{180.2 \text{ g}} = 1.15 \times 10^{22} \text{ molecules } C_6H_{12}O_6$$

b.
$$3.45 \text{ mol} \times \frac{6.022 \times 10^{23} \text{ molecules}}{1 \text{ mol}} = 2.08 \times 10^{24} \text{ molecules } C_6H_{12}O_6$$

c. molar mass of ICl_5 = 304.2 g

$$25.0 \text{ g} \times \frac{6.022 \times 10^{23} \text{ molecules}}{304.2 \text{ g}} = 4.95 \times 10^{22} \text{ molecules } ICl_5$$

d. molar mass of B_2H_6 = 27.67 g

$$1.00 \text{ g} \times \frac{6.022 \times 10^{23} \text{ molecules}}{27.67 \text{ g}} = 2.18 \times 10^{22} \text{ molecules } B_2H_6$$

e. 1.05 millimol = 0.00105 mol

$$0.00105 \text{ mol} \times \frac{6.022 \times 10^{23} \text{ molecules}}{1 \text{ mol}} = 6.32 \times 10^{20} \text{ formula units}$$

118. a. mass of 3 mol Ca = 3(40.08 g) = 120.24 g
         mass of 2 mol P = 2(30.97 g) = 61.94 g
         mass of 8 mol O = 8(16.00 g) = 128.00 g
         ───
         molar mass of $Ca_3(PO_4)_2$ =   310.18 g

   % Ca = $\dfrac{120.24 \text{ g Ca}}{310.18 \text{ g}}$ × 100 = 38.76% Ca = 38.8% Ca

   % P = $\dfrac{61.94 \text{ g P}}{310.18 \text{ g}}$ × 100 = 19.97% P = 20.0% P

   % O = $\dfrac{128.00 \text{ g O}}{310.18 \text{ g}}$ × 100 = 41.27% O = 41.3% O

   b. mass of 1 mol Cd =   112.4 g =   112.4 g
       mass of 1 mol S =   32.07 g =   32.07 g
       mass of 4 mol O = 4(16.00 g) = 64.00 g
       ───
       molar mass of $CdSO_4$ =   208.5 g

   % Cd = $\dfrac{112.4 \text{ g Cd}}{208.5 \text{ g}}$ × 100 = 53.91% Cd = 53.9% Cd

   % S = $\dfrac{32.07 \text{ g S}}{208.5 \text{ g}}$ × 100 = 15.38% S = 15.4% S

   % O = $\dfrac{64.00 \text{ g O}}{208.5 \text{ g}}$ × 100 = 30.70% O = 30.7% O

   c. mass of 2 mol Fe = 2(55.85 g) = 111.7 g
       mass of 3 mol S   = 3(32.07 g) = 96.21 g
       mass of 12 mol O = 12(16.00 g) = 192.0 g
       ───
       molar mass of $Fe_2(SO_4)_3$ =   399.9 g

   % Fe = $\dfrac{111.7 \text{ g Fe}}{399.9 \text{ g}}$ × 100 = 27.93% Fe = 27.9% Fe

   % S = $\dfrac{96.21 \text{ g S}}{399.9 \text{ g}}$ × 100 = 24.06% S = 24.1% S

   % O = $\dfrac{192.0 \text{ g O}}{399.9 \text{ g}}$ × 100 = 48.01% O = 48.0% O

   d. mass of 1 mol Mn =   54.94 g =   54.94 g
       mass of 2 mol Cl = 2(35.45 g) = 70.90 g
       ───
       molar mass of $MnCl_2$ =   125.84 g

98 Chapter 9 Chemical Composition

$$\% \text{ Mn} = \frac{54.94 \text{ g Mn}}{125.84 \text{ g}} \times 100 = 43.66\% \text{ Mn} = 43.7\% \text{ Mn}$$

$$\% \text{ Cl} = \frac{70.90 \text{ g Cl}}{125.84 \text{ g}} \times 100 = 56.34\% \text{ Cl} = 56.3\% \text{ Cl}$$

e. mass of 2 mol N = 2(14.01 g) = 28.02 g
    mass of 8 mol H = 8(1.008 g) = 8.064 g
    mass of 1 mol C = 12.01 g = 12.01 g
    mass of 3 mol O = 3(16.00 g) = 48.00 g
    ———————————————————————————————
    molar mass of $(NH_4)_2CO_3$ = 96.09 g

$$\% \text{ N} = \frac{28.02 \text{ g N}}{96.09 \text{ g}} \times 100 = 29.16\% \text{ N} = 29.2\% \text{ N}$$

$$\% \text{ H} = \frac{8.064 \text{ g H}}{96.09 \text{ g}} \times 100 = 8.392\% \text{ H} = 8.39\% \text{ H}$$

$$\% \text{ C} = \frac{12.01 \text{ g C}}{96.09 \text{ g}} \times 100 = 12.50\% \text{ C} = 12.5\% \text{ C}$$

$$\% \text{ O} = \frac{48.00 \text{ g O}}{96.09 \text{ g}} \times 100 = 49.95\% \text{ O} = 50.0\% \text{ O}$$

f. mass of 1 mol Na = 22.99 g = 22.99 g
    mass of 1 mol H = 1.008 g = 1.008 g
    mass of 1 mol C = 12.01 g = 12.01 g
    mass of 3 mol O = 3(16.00 g) = 48.00 g
    ———————————————————————————————
    molar mass of $NaHCO_3$ = 84.01 g

$$\% \text{ Na} = \frac{22.99 \text{ g Na}}{84.01 \text{ g}} \times 100 = 27.37\% \text{ Na} = 27.3\% \text{ Na}$$

$$\% \text{ H} = \frac{1.008 \text{ g H}}{84.01 \text{ g}} \times 100 = 1.200\% \text{ H} = 1.20\% \text{ H}$$

$$\% \text{ C} = \frac{12.01 \text{ g C}}{84.01 \text{ g}} \times 100 = 14.30\% \text{ C} = 14.3\% \text{ C}$$

$$\% \text{ O} = \frac{48.00 \text{ g O}}{84.01 \text{ g}} \times 100 = 57.14\% \text{ O} = 57.1\% \text{ O}$$

g. mass of 1 mol C = 12.01 g = 12.01 g
    mass of 2 mol O = 2(16.00 g) = 32.00 g
    ———————————————————————————————
    molar mass of CO_2 = 44.01 g

$$\% \text{ C} = \frac{12.01 \text{ g C}}{44.01 \text{ g}} \times 100 = 27.29\% \text{ C} = 27.3\% \text{ C}$$

$$\% \text{ O} = \frac{32.00 \text{ g O}}{44.01 \text{ g}} \times 100 = 72.71\% \text{ O} = 72.7\% \text{ O}$$

h. mass of 1 mol Ag = 107.9 g = 107.9 g
    mass of 1 mol N = 14.01 g = 14.01 g
    mass of 3 mol O = 3(16.00 g) = 48.00 g
    ─────────────────────────────────────
    molar mass of $AgNO_3$ = 169.9 g

$$\% \text{ Ag} = \frac{107.9 \text{ g Ag}}{169.9 \text{ g}} \times 100 = 63.51\% \text{ Ag} = 63.5\% \text{ Ag}$$

$$\% \text{ N} = \frac{14.01 \text{ g N}}{169.9 \text{ g}} \times 100 = 8.246\% \text{ N} = 8.25\% \text{ N}$$

$$\% \text{ O} = \frac{48.00 \text{ g O}}{169.9 \text{ g}} \times 100 = 28.25\% \text{ O} = 28.3\% \text{ O}$$

120. a. $\% \text{ Fe} = \dfrac{55.85 \text{ g Fe}}{151.91 \text{ g}} \times 100 = 36.77\% \text{ Fe}$

b. $\% \text{ Ag} = \dfrac{215.8 \text{ g Ag}}{231.8 \text{ g}} \times 100 = 93.10\% \text{ Ag}$

c. $\% \text{ Sr} = \dfrac{87.62 \text{ g Sr}}{158.5 \text{ g}} \times 100 = 55.28\% \text{ Sr}$

d. $\% \text{ C} = \dfrac{48.04 \text{ g C}}{86.08 \text{ g}} \times 100 = 55.81\% \text{ C}$

e. $\% \text{ C} = \dfrac{12.01 \text{ g C}}{32.04 \text{ g}} \times 100 = 37.48\% \text{ C}$

f. $\% \text{ Al} = \dfrac{53.96 \text{ g Al}}{101.96 \text{ g}} \times 100 = 52.92\% \text{ Al}$

g. $\% \text{ K} = \dfrac{39.10 \text{ g K}}{106.55 \text{ g}} \times 100 = 36.70\% \text{ K}$

h. $\% \text{ K} = \dfrac{39.10 \text{ g K}}{74.55 \text{ g}} \times 100 = 52.45\% \text{ K}$

122. $0.2990 \text{ g C} \times \dfrac{1 \text{ mol C}}{12.01 \text{ g C}} = 0.02490 \text{ mol C}$

$$0.05849 \text{ g H} \times \frac{1 \text{ mol H}}{1.008 \text{ g H}} = 0.05803 \text{ mol H}$$

$$0.2318 \text{ g N} \times \frac{1 \text{ mol N}}{14.01 \text{ g N}} = 0.01655 \text{ mol N}$$

$$0.1328 \text{ g O} \times \frac{1 \text{ mol O}}{16.00 \text{ g O}} = 0.008300 \text{ mol O}$$

Dividing each number of moles by the smallest number of moles (0.008300 mol O) gives

$$\frac{0.02490 \text{ mol C}}{0.008300} = 3.000 \text{ mol C}$$

$$\frac{0.05803 \text{ mol H}}{0.008300} = 6.992 \text{ mol H}$$

$$\frac{0.01655 \text{ mol N}}{0.008300} = 1.993 \text{ mol N}$$

$$\frac{0.008300 \text{ mol O}}{0.008300} = 1.000 \text{ mol O}$$

The empirical formula is $C_3H_7N_2O$.

124. Mass of oxygen in compound = 4.33 g − 4.01 g = 0.32 g O

$$4.01 \text{ g Hg} \times \frac{1 \text{ mol Hg}}{200.6 \text{ g Hg}} = 0.0200 \text{ mol Hg}$$

$$0.32 \text{ g O} \times \frac{1 \text{ mol O}}{16.00 \text{ g O}} = 0.020 \text{ mol O}$$

Since the numbers of moles are equal, the empirical formula is HgO.

126. Assume we have 100.0 g of the compound.

$$65.95 \text{ g Ba} \times \frac{1 \text{ mol Ba}}{137.3 \text{ g Ba}} = 0.4803 \text{ mol Ba}$$

$$34.05 \text{ g Cl} \times \frac{1 \text{ mol Cl}}{35.45 \text{ g Cl}} = 0.9605 \text{ mol Cl}$$

Dividing each of these number of moles by the smaller number gives

$$\frac{0.4803 \text{ mol Ba}}{0.4803} = 1.000 \text{ mol Ba}$$

$$\frac{0.9605 \text{ mol Cl}}{0.4803} = 2.000 \text{ mol Cl}$$

The empirical formula is then $BaCl_2$.

Chapter 10   Chemical Quantities

2. The coefficients of the balanced chemical equation for a reaction indicate the *relative numbers of moles* of each reactant that combine during the process, as well as the number of moles of each product formed.

4. Balanced chemical equations tell us in what proportions *on a mole basis* substances combine; since the molar masses of $C(s)$ and $O_2(g)$ are different, 1 g of O_2 could not represent the same number of moles as 1 g of C.

6. a. $3MnO_2(s) + 4Al(s) \rightarrow 3Mn(s) + 2Al_2O_3(s)$

   Three formula units of manganese(IV) oxide react with four aluminum atoms, producing three manganese atoms and two formula units of aluminum oxide. Three moles of solid manganese(IV) oxide react with four moles of solid aluminum, to produce three moles of solid manganese and two moles of solid aluminum oxide.

   b. $B_2O_3(s) + 3CaF_2(s) \rightarrow 2BF_3(g) + 3CaO(s)$

   One diboron trioxide molecule reacts with three formula units of calcium fluoride, producing two molecules of boron trifluoride and three formula units of calcium oxide. One mole of solid diboron trioxide reacts with three moles of solid calcium fluoride, to give two moles of gaseous boron trifluoride and three moles of solid calcium oxide.

   c. $3NO_2(g) + H_2O(l) \rightarrow 2HNO_3(aq) + NO(g)$

   Three molecules of nitrogen dioxide [nitrogen(IV) oxide] react with one molecule of water, to produce two molecules of nitric acid and one molecule of nitrogen monoxide [nitrogen(II) oxide]. Three moles of gaseous nitrogen dioxide react with one mole of liquid water, to produce two moles of aqueous nitric acid and one mole of nitrogen monoxide gas.

   d. $C_6H_6(g) + 3H_2(g) \rightarrow C_6H_{12}(g)$

   One molecule of C_6H_6 (which is named benzene) reacts with three molecules of hydrogen, producing just one molecule of C_6H_{12} (which is named cyclohexane). One mole of gaseous benzene reacts with three moles of hydrogen gas, giving one mole of gaseous cyclohexane.

8. False. Reactions take place on a *mole* basis: one *mole* of nitrogen gas would react with three *moles* of iodine, giving two *moles* of nitrogen triiodide.

10. $CH_3CH_2OH(l) + 3O_2(g) \rightarrow 2CO_2(g) + 3H_2O(g)$

    $$\frac{2 \text{ mol } CO_2}{1 \text{ mol } CH_3CH_2OH} \quad \text{and} \quad \frac{3 \text{ mol } H_2O}{1 \text{ mol } CH_3CH_2OH}$$

101

102   Chapter 10   Chemical Quantities

12. a. $3PbO(s) + 2Al(s) \rightarrow 3Pb(s) + Al_2O_3(s)$

$$0.25 \text{ mol PbO} \times \frac{3 \text{ mol Pb}}{3 \text{ mol PbO}} = 0.25 \text{ mol PbO}$$

$$0.25 \text{ mol PbO} \times \frac{1 \text{ mol Al}_2O_3}{3 \text{ mol PbO}} = 0.083 \text{ mol Al}_2O_3$$

b. $C_2H_2(g) + 2H_2(g) \rightarrow C_2H_6(g)$

$$0.25 \text{ mol } C_2H_2 \times \frac{1 \text{ mol } C_2H_6}{1 \text{ mol } C_2H_2} = 0.25 \text{ mol } C_2H_6$$

c. $BaO_2(s) + 2Na(s) \rightarrow Ba(s) + Na_2O_2(s)$

$$0.25 \text{ mol Na} \times \frac{1 \text{ mol Na}_2O_2}{2 \text{ mol Na}} = 0.125 = 0.13 \text{ mol Na}_2O_2$$

$$0.25 \text{ mol Na} \times \frac{1 \text{ mol Ba}}{2 \text{ mol Na}} = 0.125 = 0.13 \text{ mol Ba}$$

d. $2CH_4(g) + 3O_2(g) \rightarrow 2CO(g) + 4H_2O(l)$

$$0.25 \text{ mol CH}_4 \times \frac{2 \text{ mol CO}}{2 \text{ mol CH}_4} = 0.25 \text{ mol CO}$$

$$0.25 \text{ mol CH}_4 \times \frac{4 \text{ mol H}_2O}{2 \text{ mol CH}_4} = 0.50 \text{ mol H}_2O$$

14. a. $NH_3(g) + HCl(g) \rightarrow NH_4Cl(s)$

molar mass of NH_4Cl, 53.49 g

$$0.50 \text{ mol NH}_3 \times \frac{1 \text{ mol NH}_4Cl}{1 \text{ mol NH}_3} = 0.50 \text{ mol NH}_4Cl$$

$$0.50 \text{ mol NH}_4Cl \times \frac{53.49 \text{ g NH}_4Cl}{1 \text{ mol NH}_4Cl} = 27 \text{ g NH}_4Cl$$

b. $CH_4(g) + 4S(g) \rightarrow CS_2(l) + 2H_2S(g)$

molar masses: CS_2, 76.15 g; H_2S, 34.09 g

$$0.50 \text{ mol S} \times \frac{1 \text{ mol CS}_2}{4 \text{ mol S}} = 0.125 \text{ mol CS}_2 \ (= 0.13 \text{ mol CS}_2)$$

$$0.125 \text{ mol CS}_2 \times \frac{76.15 \text{ g CS}_2}{1 \text{ mol CS}_2} = 9.5 \text{ g CS}_2$$

$$0.50 \text{ mol S} \times \frac{2 \text{ mol H}_2S}{4 \text{ mol S}} = 0.25 \text{ mol H}_2S$$

$$0.25 \text{ mol H}_2S \times \frac{34.09 \text{ g H}_2S}{1 \text{ mol H}_2S} = 8.5 \text{ g H}_2S$$

c. $PCl_3(l) + 3H_2O(l) \rightarrow H_3PO_3(aq) + 3HCl(aq)$

molar masses: H_3PO_3, 81.99 g; HCl, 36.46 g

$$0.50 \text{ mol } PCl_3 \times \frac{1 \text{ mol } H_3PO_3}{1 \text{ mol } PCl_3} = 0.50 \text{ mol } H_3PO_3$$

$$0.50 \text{ mol } H_3PO_3 \times \frac{81.99 \text{ g } H_3PO_3}{1 \text{ mol } H_3PO_3} = 41 \text{ g } H_3PO_3$$

$$0.50 \text{ mol } PCl_3 \times \frac{3 \text{ mol } HCl}{1 \text{ mol } H_3PO_3} = 1.5 \text{ mol } HCl$$

$$1.5 \text{ mol } HCl \times \frac{36.46 \text{ g } HCl}{1 \text{ mol } HCl} = 54.7 = 55 \text{ g } HCl$$

d. $NaOH(s) + CO_2(g) \rightarrow NaHCO_3$

molar mass of $NaHCO_3$ = 84.01 g

$$0.50 \text{ mol } NaOH \times \frac{1 \text{ mol } NaHCO_3}{1 \text{ mol } NaOH} = 0.50 \text{ mol } NaHCO_3$$

$$0.50 \text{ mol } NaHCO_3 \times \frac{84.01 \text{ g } NaHCO_3}{1 \text{ mol } NaHCO_3} = 42 \text{ g } NaHCO_3$$

16. Before doing the calculations, the equations must be *balanced*.

a. $4KO_2(s) + 2H_2O(l) \rightarrow 3O_2(g) + 4KOH(s)$

$$0.625 \text{ mol } KOH \times \frac{3 \text{ mol } O_2}{4 \text{ mol } KOH} = 0.469 \text{ mol } O_2$$

b. $SeO_2(g) + 2H_2Se(g) \rightarrow 3Se(s) + 2H_2O(g)$

$$0.625 \text{ mol } H_2O \times \frac{3 \text{ mol } Se}{2 \text{ mol } H_2O} = 0.938 \text{ mol } Se$$

c. $2CH_3CH_2OH(l) + O_2(g) \rightarrow 2CH_3CHO(aq) + 2H_2O(l)$

$$0.625 \text{ mol } H_2O \times \frac{2 \text{ mol } CH_3CHO}{2 \text{ mol } H_2O} = 0.625 \text{ mol } CH_3CHO$$

d. $Fe_2O_3(s) + 2Al(s) \rightarrow 2Fe(l) + Al_2O_3(s)$

$$0.625 \text{ mol } Al_2O_3 \times \frac{2 \text{ mol } Fe}{1 \text{ mol } Al_2O_3} = 1.25 \text{ mol } Fe$$

18. Stoichiometry is the process of using a chemical equation to calculate the relative masses of reactants and products involved in a reaction.

20. a. molar mass of Ag = 107.9 g

$$2.01 \times 10^{-2} \text{ g Ag} \times \frac{1 \text{ mol}}{107.9 \text{ g}} = 1.86 \times 10^{-4} \text{ mol Ag}$$

104 Chapter 10 Chemical Quantities

b. molar mass of $(NH_4)_2S$ = 68.15 g

$$45.2 \text{ mg } (NH_4)_2S \times \frac{1 \text{ g}}{1000 \text{ mg}} \times \frac{1 \text{ mol}}{68.15 \text{ g}} = 6.63 \times 10^{-4} \text{ mol } (NH_4)_2S$$

c. molar mass of uranium = 238.0 g

$$61.7 \text{ } \mu g \text{ U} \times \frac{1 \text{ g}}{10^6 \text{ } \mu g} \times \frac{1 \text{ mol}}{238.0 \text{ g}} = 2.59 \times 10^{-7} \text{ mol U}$$

d. molar mass of SO_2 = 64.07 g

$$5.23 \text{ kg } SO_2 \times \frac{1000 \text{ g}}{1 \text{ kg}} \times \frac{1 \text{ mol}}{64.07 \text{ g}} = 81.6 \text{ mol } SO_2$$

e. molar mass of $Fe(NO_3)_3$ = 241.9 g

$$272 \text{ g } Fe(NO_3)_3 \times \frac{1 \text{ mol}}{241.9 \text{ g}} = 1.12 \text{ mol } Fe(NO_3)_3$$

f. molar mass of $FeSO_4$ = 151.9 g

$$12.7 \text{ mg } FeSO_4 \times \frac{1 \text{ g}}{1000 \text{ mg}} \times \frac{1 \text{ mol}}{151.9 \text{ g}} = 8.36 \times 10^{-5} \text{ mol } FeSO_4$$

g. molar mass of LiOH = 23.95 g

$$6.91 \times 10^3 \text{ g LiOH} \times \frac{1 \text{ mol}}{23.95 \text{ g}} = 288.5 = 289 \text{ mol LiOH}$$

22. a. molar mass of $(C_2H_5)_2O$ = 74.12 g

$$0.00249 \text{ mol } (C_2H_5)_2O \times \frac{74.12 \text{ g}}{1 \text{ mol}} = 0.185 \text{ g } (C_2H_5)_2O$$

b. molar mass of UF_6 = 352.0 g

$$1.37 \text{ mol } UF_6 \times \frac{352.0 \text{ g}}{1 \text{ mol}} = 482 \text{ g } UF_6$$

c. molar mass of SO_3 = 80.07 g

$$3.65 \times 10^{-5} \text{ mol } SO_3 \times \frac{80.07 \text{ g}}{1 \text{ mol}} = 2.92 \times 10^{-3} \text{ g } SO_3$$

d. molar mass of K_2CO_3 = 138.2 g

$$52.6 \text{ mol } K_2CO_3 \times \frac{138.2 \text{ g}}{1 \text{ mol}} = 7.27 \times 10^3 \text{ g } K_2CO_3$$

e. molar mass of NaOH = 40.00 g

$$2.06 \times 10^4 \text{ mol NaOH} \times \frac{40.00 \text{ g}}{1 \text{ mol}} = 8.24 \times 10^5 \text{ g NaOH}$$

f. molar mass of $CuSO_4$ = 159.6 g

$$71.5 \text{ millimol} \times \frac{1 \text{ mol}}{10^3 \text{ millimol}} \times \frac{159.6 \text{ g}}{1 \text{ mol}} = 11.4 \text{ g } CuSO_4$$

g. molar mass of $C_6H_4Cl_2$ = 147.0 g

$$3.37 \text{ mol } C_6H_4Cl_2 \times \frac{147.0 \text{ g}}{1 \text{ mol}} = 495 \text{ g } C_6H_4Cl_2$$

24. Before any calculations are done, the equations must be *balanced*.

  a. $Mg(s) + CuCl_2(aq) \rightarrow MgCl_2(aq) + Cu(s)$

   molar mass of Mg = 24.31 g

   $$25.0 \text{ g Mg} \times \frac{1 \text{ mol}}{24.31 \text{ g}} = 1.03 \text{ mol Mg}$$

   $$1.03 \text{ mol Mg} \times \frac{1 \text{ mol } CuCl_2}{1 \text{ mol Mg}} = 1.03 \text{ mol } CuCl_2$$

  b. $2AgNO_3(aq) + NiCl_2(aq) \rightarrow 2AgCl(s) + Ni(NO_3)_2(aq)$

   molar mass of $AgNO_3$ = 169.9 g

   $$25.0 \text{ g } AgNO_3 \times \frac{1 \text{ mol}}{169.9 \text{ g}} = 0.147 \text{ mol } AgNO_3$$

   $$0.147 \text{ mol } AgNO_3 \times \frac{1 \text{ mol } NiCl_2}{2 \text{ mol } AgNO_3} = 0.0735 \text{ mol } NiCl_2$$

  c. $NaHSO_3(aq) + NaOH(aq) \rightarrow Na_2SO_3(aq) + H_2O(l)$

   molar mass of $NaHSO_3$ = 104.1 g

   $$25.0 \text{ g } NaHSO_3 \times \frac{1 \text{ mol}}{104.1 \text{ g}} = 0.240 \text{ mol } NaHSO_3$$

   $$0.240 \text{ mol } NaHSO_3 \times \frac{1 \text{ mol NaOH}}{1 \text{ mol } NaHSO_3} = 0.240 \text{ mol NaOH}$$

  d. $KHCO_3(aq) + HCl(aq) \rightarrow KCl(aq) + H_2O(l) + CO_2(g)$

   molar mass of $KHCO_3$ = 100.1 g

   $$25.0 \text{ g } KHCO_3 \times \frac{1 \text{ mol}}{100.1 \text{ g}} = 0.250 \text{ mol } KHCO_3$$

   $$0.250 \text{ mol } KHCO_3 \times \frac{1 \text{ mol HCl}}{1 \text{ mol } KHCO_3} = 0.250 \text{ mol HCl}$$

26. Before any calculations are done, the equations must be *balanced*. Since the given and required quantities in this question are given in *milligrams*, it is most convenient to perform the calculations in terms of *millimoles* of the substances involved. One millimole of a substance represents the molar mass of the substance expressed in milligrams.

106   Chapter 10   Chemical Quantities

a. $FeSO_4(aq) + K_2CO_3(aq) \rightarrow FeCO_3(s) + K_2SO_4(aq)$

millimolar masses: $FeSO_4$, 151.9 mg; $FeCO_3$, 115.9 mg; K_2SO_4, 174.3 mg

$$10.0 \text{ mg} \times \frac{1 \text{ mmol FeSO}_4}{151.9 \text{ mg FeSO}_4} = 0.0658 \text{ mmol FeSO}_4$$

$$0.0658 \text{ mmol FeSO}_4 \times \frac{1 \text{ mmol FeCO}_3}{1 \text{ mmol FeSO}_4} \times \frac{115.9 \text{ mg FeCO}_3}{1 \text{ mmol FeCO}_3} = 7.63 \text{ mg FeCO}_3$$

$$0.0658 \text{ mmol FeSO}_4 \times \frac{1 \text{ mmol K}_2SO_4}{1 \text{ mmol FeSO}_4} \times \frac{174.3 \text{ mg K}_2SO_4}{1 \text{ mmol K}_2SO_4} = 11.5 \text{ mg K}_2SO_4$$

b. $4Cr(s) + 3SnCl_4(l) \rightarrow 4CrCl_3(s) + 3Sn(s)$

millimolar masses: Cr, 52.00 mg; $CrCl_3$, 158.4 mg; Sn, 118.7 mg

$$10.0 \text{ mg Cr} \times \frac{1 \text{ mmol Cr}}{52.00 \text{ mg Cr}} = 0.192 \text{ mmol Cr}$$

$$0.192 \text{ mmol Cr} \times \frac{4 \text{ mmol CrCl}_3}{4 \text{ mmol Cr}} \times \frac{158.4 \text{ mg CrCl}_3}{1 \text{ mmol CrCl}_3} = 30.4 \text{ mg CrCl}_3$$

$$0.192 \text{ mmol Cr} \times \frac{3 \text{ mmol Sn}}{4 \text{ mmol Cr}} \times \frac{118.7 \text{ mg Sn}}{1 \text{ mmol Sn}} = 17.1 \text{ mg Sn}$$

c. $16Fe(s) + 3S_8(s) \rightarrow 8Fe_2S_3(s)$

millimolar masses: S_8, 256.6 mg; Fe_2S_3, 207.9 mg

$$10.0 \text{ mg S}_8 \times \frac{1 \text{ mmol S}_8}{256.6 \text{ mg S}_8} = 0.0390 \text{ mmol S}_8$$

$$0.0390 \text{ mmol S}_8 \times \frac{8 \text{ mmol Fe}_2S_3}{3 \text{ mmol S}_8} \times \frac{207.9 \text{ mg Fe}_2S_3}{1 \text{ mmol Fe}_2S_3} = 21.6 \text{ mg Fe}_2S_3$$

d. $3Ag(s) + 4HNO_3(aq) \rightarrow 3AgNO_3(aq) + 2H_2O(l) + NO(g)$

millimolar masses: HNO_3, 63.0 mg; $AgNO_3$, 169.9 mg
H_2O, 18.0 mg;   NO, 30.0 mg

$$10.0 \text{ mg HNO}_3 \times \frac{1 \text{ mmol HNO}_3}{63.0 \text{ mg HNO}_3} = 0.159 \text{ mmol HNO}_3$$

$$0.159 \text{ mmol HNO}_3 \times \frac{3 \text{ mmol AgNO}_3}{4 \text{ mmol HNO}_3} \times \frac{169.9 \text{ mg AgNO}_3}{1 \text{ mmol AgNO}_3} = 20.3 \text{ mg AgNO}_3$$

$$0.159 \text{ mmol HNO}_3 \times \frac{2 \text{ mmol H}_2O}{4 \text{ mmol HNO}_3} \times \frac{18.0 \text{ mg H}_2O}{1 \text{ mmol H}_2O} = 1.43 \text{ mg H}_2O$$

$$0.159 \text{ mmol HNO}_3 \times \frac{1 \text{ mmol NO}}{4 \text{ mmol HNO}_3} \times \frac{30.0 \text{ mg NO}}{1 \text{ mmol NO}} = 1.19 \text{ mg NO}$$

Chapter 10   Chemical Quantities   107

28. $2H_2(g) + O_2(g) \rightarrow 2H_2O(g)$

molar masses: H_2, 2.016 g; O_2, 32.00 g

$$1.00 \text{ g } H_2 \times \frac{1 \text{ mol } H_2}{2.016 \text{ g } H_2} = 0.496 \text{ mol } H_2$$

$$0.496 \text{ mol } H_2 \times \frac{1 \text{ mol } O_2}{2 \text{ mol } H_2} = 0.248 \text{ mol } O_2$$

$$0.248 \text{ mol } O_2 \times \frac{32.00 \text{ g } O_2}{1 \text{ mol } O_2} = 7.94 \text{ g } O_2$$

30. $2NI_3(s) \rightarrow N_2(g) + 3I_2(g)$

molar masses: NI_3, 394.7 g; I_2, 253.8 g

$$1.75 \text{ g } NI_3 \times \frac{1 \text{ mol } NI_3}{394.7 \text{ g } NI_3} = 4.43 \times 10^{-3} \text{ mol } NI_3$$

$$4.43 \times 10^{-3} \text{ mol } NI_3 \times \frac{3 \text{ mol } I_2}{2 \text{ mol } NI_3} = 6.65 \times 10^{-3} \text{ mol } I_2$$

$$6.65 \times 10^{-3} \text{ mol } I_2 \times \frac{253.8 \text{ g } I_2}{1 \text{ mol } I_2} = 1.69 \text{ g } I_2$$

32. $Zn(s) + 2HCl(aq) \rightarrow ZnCl_2(aq) + H_2(g)$

$2Al(s) + 6HCl(aq) \rightarrow 2AlCl_3(aq) + 3H_2(g)$

$Mg(s) + 2HCl(aq) \rightarrow MgCl_2(aq) + H_2(g)$

molar masses: H_2, 2.016 g; Zn, 65.38 g; Al, 26.98 g; Mg, 24.31 g

$$1.00 \text{ g } H_2 \times \frac{1 \text{ mol}}{2.016 \text{ g}} = 0.496 \text{ mol } H_2$$

$$0.496 \text{ mol } H_2 \times \frac{1 \text{ mol Zn}}{1 \text{ mol } H_2} \times \frac{65.38 \text{ g Zn}}{1 \text{ mol Zn}} = 32.4 \text{ g Zn}$$

$$0.496 \text{ mol } H_2 \times \frac{2 \text{ mol Al}}{3 \text{ mol } H_2} \times \frac{26.98 \text{ g Al}}{1 \text{ mol Al}} = 8.92 \text{ g Al}$$

$$0.496 \text{ mol } H_2 \times \frac{1 \text{ mol Mg}}{1 \text{ mol } H_2} \times \frac{24.31 \text{ g Mg}}{1 \text{ mol Mg}} = 12.1 \text{ g Mg}$$

34. $Cl_2(g) + F_2(g) \rightarrow 2ClF(g)$

molar masses: Cl_2, 70.90 g; F_2, 38.00 g

$$5.00 \text{ mg } F_2 \times \frac{1 \text{ g}}{10^3 \text{ mg}} \times \frac{1 \text{ mol } F_2}{38.00 \text{ g } F_2} = 1.315 \times 10^{-4} \text{ mol } F_2$$

108   Chapter 10   Chemical Quantities

$$1.315 \times 10^{-4} \text{ mol } F_2 \times \frac{1 \text{ mol } Cl_2}{1 \text{ mol } F_2} \times \frac{70.90 \text{ g } Cl_2}{1 \text{ mol } Cl_2} = 9.33 \times 10^{-3} \text{ g } Cl_2$$

36. $Hg(l) + S(s) \rightarrow HgS(s)$

    molar masses: Hg, 200.6 g; S, 32.07 g

    $$0.10 \text{ g Hg} \times \frac{1 \text{ mol}}{200.6 \text{ g}} = 4.99 \times 10^{-4} \text{ mol Hg}$$

    $$4.99 \times 10^{-4} \text{ mol Hg} \times \frac{1 \text{ mol S}}{1 \text{ mol Hg}} = 4.99 \times 10^{-4} \text{ mol S}$$

    $$4.99 \times 10^{-4} \text{ mol S} \times \frac{32.07 \text{ g}}{1 \text{ mol}} = 1.60 \times 10^{-2} \text{ g S}$$

38. $2NaOH(aq) + CO_2(g) \rightarrow Na_2CO_3(aq) + H_2O(l)$

    molar masses: NaOH, 40.00 g; CO_2, 44.01 g

    $$5.00 \text{ g NaOH} \times \frac{1 \text{ mol}}{40.00 \text{ g}} = 0.125 \text{ mol NaOH}$$

    $$0.125 \text{ mol NaOH} \times \frac{1 \text{ mol } CO_2}{2 \text{ mol NaOH}} = 6.25 \times 10^{-2} \text{ mol } CO_2$$

    $$6.25 \times 10^{-2} \text{ mol } CO_2 \times \frac{44.01 \text{ g}}{1 \text{ mol}} = 2.75 \text{ g } CO_2$$

40. $2Mg(s) + O_2(g) \rightarrow 2MgO(s)$

    molar masses: Mg, 24.31 g; MgO, 40.31 g

    $$1.25 \text{ g Mg} \times \frac{1 \text{ mol}}{24.31 \text{ g}} = 5.14 \times 10^{-2} \text{ mol Mg}$$

    $$5.14 \times 10^{-2} \text{ mol Mg} \times \frac{2 \text{ mol MgO}}{2 \text{ mol Mg}} = 5.14 \times 10^{-2} \text{ mol MgO}$$

    $$5.14 \times 10^{-2} \text{ mol MgO} \times \frac{40.31 \text{ g}}{1 \text{ mol}} = 2.07 \text{ g MgO}$$

42. To determine the limiting reactant, first calculate the number of moles of each reactant present. Then determine how these numbers of moles correspond to the stoichiometric ratio indicated by the balanced chemical equation for the reaction.

44. A reactant is present *in excess* if there is more of that reactant present than is needed to combine with the limiting reactant for the process. An excess of any reactant does not affect the theoretical yield for a process: the theoretical yield is determined by the limiting reactant.

Chapter 10 Chemical Quantities 109

46. **a.** $2Al(s) + 6HCl(aq) \rightarrow 2AlCl_3(aq) + 3H_2(g)$

Molar masses: Al, 26.98 g; HCl, 36.46 g; $AlCl_3$, 133.3 g; H_2, 2.016 g

$$15.0 \text{ g Al} \times \frac{1 \text{ mol}}{26.98 \text{ g}} = 0.556 \text{ mol Al}$$

$$15.0 \text{ g HCl} \times \frac{1 \text{ mol}}{36.46 \text{ g}} = 0.411 \text{ mol HCl}$$

Since HCl is needed to react with Al in a 6:2 (i.e., 3:1) molar ratio, it seems pretty certain that HCl is the limiting reactant. To prove this, we can calculate the quantity of Al that would react with the given number of moles of HCl.

$$0.411 \text{ mol HCl} \times \frac{2 \text{ mol Al}}{6 \text{ mol HCl}} = 0.137 \text{ mol Al}$$

By this calculation we have shown that *all* the HCl present will be needed to react with only 0.137 mol Al (out of the 0.556 mol Al present). Therefore HCl is the limiting reactant, and Al is present in excess. The calculation of the masses of products produced is based on the number of moles of the limiting reactant.

$$0.411 \text{ mol HCl} \times \frac{2 \text{ mol AlCl}_3}{6 \text{ mol HCl}} \times \frac{133.3 \text{ g}}{1 \text{ mol}} = 18.3 \text{ g AlCl}_3$$

$$0.411 \text{ mol HCl} \times \frac{3 \text{ mol H}_2}{6 \text{ mol HCl}} \times \frac{2.016 \text{ g}}{1 \text{ mol}} = 0.414 \text{ g H}_2$$

b. $2NaOH(aq) + CO_2(g) \rightarrow Na_2CO_3(aq) + H_2O(l)$

molar masses: NaOH, 40.00 g; CO_2, 44.01 g; Na_2CO_3, 105.99 g; H_2O, 18.02 g

$$15.0 \text{ g NaOH} \times \frac{1 \text{ mol}}{40.00 \text{ g}} = 0.375 \text{ mol NaOH}$$

$$15.0 \text{ g CO}_2 \times \frac{1 \text{ mol}}{44.01 \text{ g}} = 0.341 \text{ mol CO}_2$$

For the 0.375 mol NaOH, let's calculate if there is enough CO_2 present to react:

$$0.375 \text{ mol NaOH} \times \frac{1 \text{ mol CO}_2}{2 \text{ mol NaOH}} = 0.1875 \text{ mol CO}_2 \text{ (0.188 mol)}$$

We have present *more* CO_2 (0.341 mol) than is needed to react with the given quantity of NaOH. NaOH is therefore the limiting reactant, and CO_2 is present in excess. The quantities of products resulting are based on the complete conversion of the limiting reactant (0.375 mol NaOH):

$$0.375 \text{ mol NaOH} \times \frac{1 \text{ mol Na}_2\text{CO}_3}{2 \text{ mol NaOH}} \times \frac{105.99 \text{ g}}{1 \text{ mol}} = 19.9 \text{ g Na}_2\text{CO}_3$$

$$0.375 \text{ mol NaOH} \times \frac{1 \text{ mol H}_2\text{O}}{2 \text{ mol NaOH}} \times \frac{18.02 \text{ g}}{1 \text{ mol}} = 3.38 \text{ g H}_2\text{O}$$

c. $Pb(NO_3)_2(aq) + 2HCl(aq) \rightarrow PbCl_2(s) + 2HNO_3(aq)$

Molar masses: $Pb(NO_3)_2$, 331.2 g; HCl, 36.46 g; $PbCl_2$, 278.1 g; HNO_3, 63.02 g

$$15.0 \text{ g Pb(NO}_3)_2 \times \frac{1 \text{ mol}}{331.2 \text{ g}} = 0.0453 \text{ mol Pb(NO}_3)_2$$

$$15.0 \text{ g HCl} \times \frac{1 \text{ mol}}{36.46 \text{ g}} = 0.411 \text{ mol}$$

With such a large disparity between the numbers of moles of the reactants, it's probably a sure bet that $Pb(NO_3)_2$ is the limiting reactant. To confirm this, we can calculate how many mol of HCl are needed to react with the given amount of $Pb(NO_3)_2$.

$$0.0453 \text{ mol Pb(NO}_3)_2 \times \frac{2 \text{ mol HCl}}{1 \text{ mol Pb(NO}_3)_2} = 0.0906 \text{ mol HCl}$$

We have considerably more HCl present than is needed to react completely with the $Pb(NO_3)_2$. Therefore $Pb(NO_3)_2$ is the limiting reactant, and HCl is present in excess. The quantities of products produced are based on the limiting reactant being completely consumed.

$$0.0453 \text{ mol Pb(NO}_3)_2 \times \frac{1 \text{ mol PbCl}_2}{1 \text{ mol Pb(NO}_3)_2} \times \frac{278.1 \text{ g}}{1 \text{ mol}} = 12.6 \text{ g PbCl}_2$$

$$0.0453 \text{ mol Pb(NO}_3)_2 \times \frac{2 \text{ mol HNO}_3}{1 \text{ mol Pb(NO}_3)_2} \times \frac{63.02 \text{ g}}{1 \text{ mol}} = 5.71 \text{ g HNO}_3$$

d. $2K(s) + I_2(s) \rightarrow 2KI(s)$

Molar masses: K, 39.10 g; I_2, 253.8 g; KI, 166.0 g

$$15.0 \text{ g K} \times \frac{1 \text{ mol}}{39.10 \text{ g}} = 0.384 \text{ mol K}$$

$$15.0 \text{ g I}_2 \times \frac{1 \text{ mol}}{253.8 \text{ g}} = 0.0591 \text{ mol I}_2$$

Since, from the balanced chemical equation, we need twice as many moles of K as moles of I_2, and since there is so little I_2, present, it's a safe bet that I_2 is the limiting reactant. To confirm this, we can calculate how many moles of K are needed to react with the given amount of I_2 present:

$$0.0591 \text{ mol } I_2 \times \frac{2 \text{ mol K}}{1 \text{ mol } I_2} = 0.1182 \text{ mol K}$$

Clearly we have more potassium present than is needed to react with the small amount of I_2 present. I_2 is therefore the limiting reactant, and potassium is present in excess. The amount of product produced is calculated from the number of moles of the limiting reactant present:

$$0.0591 \text{ mol } I_2 \times \frac{2 \text{ mol KI}}{1 \text{ mol } I_2} \times \frac{166.0 \text{ g}}{1 \text{ mol}} = 19.6 \text{ g KI}$$

48. a. $2NH_3(g) + 2Na(s) \rightarrow 2NaNH_2(s) + H_2(g)$

Molar masses: NH_3, 17.03 g; Na, 22.99 g; $NaNH_2$, 39.02 g

$$50.0 \text{ g } NH_3 \times \frac{1 \text{ mol}}{17.03 \text{ g}} = 2.94 \text{ mol } NH_3$$

$$50.0 \text{ g Na} \times \frac{1 \text{ mol}}{22.99 \text{ g}} = 2.17 \text{ mol Na}$$

Since the coefficients of NH_3 and Na are the *same* in the balanced chemical equation for the reaction, the two reactants combine in a 1:1 molar ratio. Therefore Na is the limiting reactant, which will control the amount of product produced.

$$2.17 \text{ mol Na} \times \frac{2 \text{ mol } NaNH_2}{2 \text{ mol Na}} \times \frac{39.02 \text{ g}}{1 \text{ mol}} = 84.7 \text{ g } NaNH_2$$

b. $BaCl_2(aq) + Na_2SO_4(aq) \rightarrow BaSO_4(s) + 2NaCl(aq)$

Molar masses: $BaCl_2$, 208.2 g; Na_2SO_4, 142.1 g; $BaSO_4$, 233.4 g

$$50.0 \text{ g } BaCl_2 \times \frac{1 \text{ mol}}{208.2 \text{ g}} = 0.240 \text{ mol } BaCl_2$$

$$50.0 \text{ g } Na_2SO_4 \times \frac{1 \text{ mol}}{142.1 \text{ g}} = 0.352 \text{ mol } Na_2SO_4$$

Since the coefficients of $BaCl_2$ and Na_2SO_4 are the *same* in the balanced chemical equation for the reaction, the reactant having the smaller number of moles present ($BaCl_2$) must be the limiting reactant, which will control the amount of product produced.

$$0.240 \text{ mol } BaCl_2 \times \frac{1 \text{ mol } BaSO_4}{1 \text{ mol } BaCl_2} \times \frac{233.4 \text{ g}}{1 \text{ mol}} = 56.0 \text{ g } BaSO_4$$

c. $SO_2(g) + 2NaOH(aq) \rightarrow Na_2SO_3(aq) + H_2O(l)$

Molar masses: SO_2, 64.07 g; NaOH, 40.00 g; Na_2SO_3, 126.1 g

$$50.0 \text{ g } SO_2 \times \frac{1 \text{ mol}}{64.07 \text{ g}} = 0.780 \text{ mol } SO_2$$

$$50.0 \text{ g NaOH} \times \frac{1 \text{ mol}}{40.00 \text{ g}} = 1.25 \text{ mol NaOH}$$

From the balanced chemical equation for the reaction, every time one mol of SO_2 reacts, two mol of NaOH are needed. For 0.780 mol of SO_2, 2(0.780 mol) = 1.56 mol of NaOH would be needed. We do not have sufficient NaOH to react with the SO_2 present: therefore, NaOH is the limiting reactant, and controls the amount of product obtained.

$$1.25 \text{ mol NaOH} \times \frac{1 \text{ mol } Na_2SO_3}{2 \text{ mol NaOH}} \times \frac{126.1 \text{ g}}{1 \text{ mol}} = 78.8 \text{ g } Na_2SO_3$$

d. $2Al(s) + 3H_2SO_4(l) \rightarrow Al_2(SO_4)_3(s) + 3H_2(g)$

Molar masses: Al, 26.98 g; H_2SO_4, 98.09 g; $Al_2(SO_4)_3$, 342.2 g

$$50.0 \text{ g Al} \times \frac{1 \text{ mol}}{26.98 \text{ g}} = 1.85 \text{ mol Al}$$

$$50.0 \text{ g } H_2SO_4 \times \frac{1 \text{ mol}}{98.09 \text{ g}} = 0.510 \text{ mol } H_2SO_4$$

Since the amount of H_2SO_4 present is smaller than the amount of Al, let's see if H_2SO_4 is the limiting reactant, by calculating how much Al would react with the given amount of H_2SO_4.

$$0.510 \text{ mol } H_2SO_4 \times \frac{2 \text{ mol Al}}{3 \text{ mol } H_2SO_4} = 0.340 \text{ mol Al}$$

Since all the H_2SO_4 present would react with only a small portion of the Al present, H_2SO_4 is therefore the limiting reactant and will control the amount of product obtained.

$$0.510 \text{ mol } H_2SO_4 \times \frac{1 \text{ mol } Al_2(SO_4)_3}{3 \text{ mol } H_2SO_4} \times \frac{342.2 \text{ g}}{1 \text{ mol}} = 58.2 \text{ g } Al_2(SO_4)_3$$

50. a. $CO(g) + 2H_2(g) \rightarrow CH_3OH(l)$

CO is the limiting reactant; 11.4 mg CH_3OH

b. $2Al(s) + 3I_2(s) \rightarrow 2AlI_3(s)$

I_2 is the limiting reactant; 10.7 mg AlI_3

c. $Ca(OH)_2(aq) + 2HBr(aq) \rightarrow CaBr_2(aq) + 2H_2O(l)$

HBr is the limiting reactant; 12.4 mg $CaBr_2$; 2.23 mg H_2O

d. $2Cr(s) + 2H_3PO_4(aq) \rightarrow 2CrPO_4(s) + 3H_2(g)$

H_3PO_4 is the limiting reactant; 15.0 mg $CrPO_4$; 0.309 mg H_2

52. $2NH_3(g) + CO_2(g) \rightarrow CN_2H_4O(s) + H_2O(l)$

molar masses: NH_3, 17.03 g; CO_2, 44.01 g; CN_2H_4O, 60.06 g

$$100. \text{ g } NH_3 \times \frac{1 \text{ mol } NH_3}{17.03 \text{ g } NH_3} = 5.872 \text{ mol } NH_3$$

$$100. \text{ g } CO_2 \times \frac{1 \text{ mol } CO_2}{44.01 \text{ g } CO_2} = 2.272 \text{ mol } CO_2$$

See if CO_2 is the limiting reactant.

$$2.272 \text{ mol } CO_2 \times \frac{2 \text{ mol } NH_3}{1 \text{ mol } CO_2} = 4.544 \text{ mol } NH_3$$

CO_2 is indeed the limiting reactant.

$$2.272 \text{ mol } CO_2 \times \frac{1 \text{ mol } CN_2H_4O}{1 \text{ mol } CO_2} = 2.272 \text{ mol } CN_2H_4O$$

$$2.272 \text{ mol } CN_2H_4O \times \frac{60.03 \text{ g } CN_2H_4O}{1 \text{ mol } CN_2H_4O} = 136 \text{ g } CN_2H_4O$$

54. $2Fe(s) + 3Cl_2(g) \rightarrow 2FeCl_3(s)$

Molar masses: Fe, 55.85 g; Cl_2, 70.90 g; $FeCl_3$, 162.3 g

$$5.00 \text{ g Fe} \times \frac{1 \text{ mol}}{55.85 \text{ g}} = 0.0895 \text{ mol Fe present}$$

$$10.0 \text{ g } Cl_2 \times \frac{1 \text{ mol}}{70.90 \text{ g}} = 0.141 \text{ mol } Cl_2 \text{ present}$$

Calculate how much Cl_2 is needed to react with given amount of Fe:

$$0.0895 \text{ mol Fe} \times \frac{3 \text{ mol } Cl_2}{1 \text{ mol Fe}} = 0.134 \text{ mol } Cl_2$$

We have more than this amount Cl_2 present, Fe is the limiting reactant.

$$0.0895 \text{ mol Fe} \times \frac{2 \text{ mol } FeCl_3}{2 \text{ mol Fe}} \times \frac{162.2 \text{ g}}{1 \text{ mol}} = 14.5 \text{ g } FeCl_3$$

Since the total mass of the reactant was 10.0 + 5.00 = 15.0 g, and since there is obtained 14.5 g of the product, there must be 0.5 g of unreacted Cl_2 still present.

114   Chapter 10   Chemical Quantities

56. $2CuSO_4(aq) + 5KI(aq) \rightarrow 2CuI(s) + KI_3(aq) + 2K_2SO_4(aq)$

Molar masses: $CuSO_4$, 159.6 g; KI, 166.0 g; CuI, 190.5 g; KI_3, 419.8 g; K_2SO_4, 174.3 g

$$0.525 \text{ g } CuSO_4 \times \frac{1 \text{ mol}}{159.6 \text{ g}} = 3.29 \times 10^{-3} \text{ mol } CuSO_4$$

$$2.00 \text{ g KI} \times \frac{1 \text{ mol}}{166.0 \text{ g}} = 0.0120 \text{ mol KI}$$

To determine the limiting reactant, let's calculate what amount of KI would be needed to react with the given amount of $CuSO_4$ present

$$3.29 \times 10^{-3} \text{ mol } CuSO_4 \times \frac{5 \text{ mol KI}}{2 \text{ mol } CuSO_4} = 8.23 \times 10^{-3} \text{ mol KI}$$

Since we have more KI present than the amount required to react with the $CuSO_4$ present, $CuSO_4$ must be the limiting reactant, which will control the amounts of products produced.

$$3.29 \times 10^{-3} \text{ mol } CuSO_4 \times \frac{2 \text{ mol CuI}}{2 \text{ mol } CuSO_4} \times \frac{190.5 \text{ g}}{1 \text{ mol}} = 0.627 \text{ g CuI}$$

$$3.29 \times 10^{-3} \text{ mol } CuSO_4 \times \frac{1 \text{ mol } KI_3}{2 \text{ mol } CuSO_4} \times \frac{419.8 \text{ g}}{1 \text{ mol}} = 0.691 \text{ g } KI_3$$

$$3.29 \times 10^{-3} \text{ mol } CuSO_4 \times \frac{2 \text{ mol } K_2SO_4}{2 \text{ mol } CuSO_4} \times \frac{174.3 \text{ g}}{1 \text{ mol}} = 0.573 \text{g } K_2SO_4$$

58. $SiO_2(s) + C(s) \rightarrow 2CO(g) + SiC(s)$

molar masses: SiO_2, 60.09 g; SiC, 40.1

1.0 kg = 1.0×10^3 g

$$1.0 \times 10^3 \text{ g } SiO_2 \times \frac{1 \text{ mol } SiO_2}{60.09 \text{ g } SiO_2} = 16.64 \text{ mol } SiO_2$$

From the balanced chemical equation, if 16.64 mol of SiO_2 were to react completely (an excess of carbon is present), then 16.64 mol of SiC should be produced (the coefficients of SiO_2 and SiC are the same).

$$16.64 \text{ mol SiC} \times \frac{40.01 \text{ g SiC}}{1 \text{ mol SiC}} = 6.7 \times 10^2 \text{ g SiC} = 0.67 \text{ kg SiC}$$

60. If the reaction is performed in a solvent, the product may have a substantial solubility in the solvent; the reaction may come to equilibrium before the full yield of product is achieved (see Chapter 16); loss of product may occur through operator error.

62. Percent yield = $\dfrac{\text{actual yield}}{\text{theoretical yield}} \times 100 = \dfrac{1.279 \text{ g}}{1.352 \text{ g}} \times 100 = 94.60\%$

64. $2LiOH(s) + CO_2(g) \rightarrow Li_2CO_3(s) + H_2O(g)$

molar masses: LiOH, 23.95 g; CO_2, 44.01 g

$$155 \text{ g LiOH} \times \frac{1 \text{ mol LiOH}}{23.95 \text{ g LiOH}} \times \frac{1 \text{ mol } CO_2}{2 \text{ mol LiOH}} \times \frac{44.01 \text{ g } CO_2}{1 \text{ mol } CO_2} = 142 \text{ g } CO_2$$

Since the cartridge has only absorbed 102 g CO_2 out of a total capacity of 142 g CO_2, the cartridge has absorbed

$$\frac{102 \text{ g}}{142 \text{ g}} \times 100 = 71.8\% \text{ of its capacity}$$

66. $CaCO_3(s) + 2HCl(g) \rightarrow CaCl_2(s) + CO_2(g) + H_2O(g)$

molar masses: $CaCO_3$, 100.1 g; HCl, 36.46 g; $CaCl_2$, 111.0 g

$$155 \text{ g } CaCO_3 \times \frac{1 \text{ mol } CaCO_3}{100.1 \text{ g } CaCO_3} = 1.548 \text{ mol } CaCO_3$$

$$250. \text{ g HCl} \times \frac{1 \text{ mol HCl}}{36.46 \text{ g HCl}} = 6.857 \text{ mol HCl}$$

$CaCO_3$ is the limiting reactant.

$$1.548 \text{ mol } CaCO_3 \times \frac{1 \text{ mol } CaCl_2}{1 \text{ mol } CaCO_3} = 1.548 \text{ mol } CaCl_2$$

$$1.548 \text{ mol } CaCl_2 \times \frac{111.0 \text{ g } CaCl_2}{1 \text{ mol } CaCl_2} = 172 \text{ g } CaCl_2$$

Percent yield = $\dfrac{\text{actual yield}}{\text{theoretical yield}} \times 100 = \dfrac{142 \text{ g}}{172 \text{ g}} \times 100 = 82.6\%$

68. $NaCl(aq) + NH_3(aq) + H_2O(l) + CO_2(s) \rightarrow NH_4Cl(aq) + NaHCO_3(s)$

molar masses: NH_3, 17.03 g; CO_2, 44.01 g; $NaHCO_3$, 84.01 g

$$10.0 \text{ g } NH_3 \times \frac{1 \text{ mol } NH_3}{17.03 \text{ g } NH_3} = 0.5872 \text{ mol } NH_3$$

$$15.0 \text{ g } CO_2 \times \frac{1 \text{ mol } CO_2}{44.01 \text{ g } CO_2} = 0.3408 \text{ mol } CO_2$$

CO_2 is the limiting reactant.

$$0.3408 \text{ mol } CO_2 \times \frac{1 \text{ mol } NaHCO_3}{1 \text{ mol } CO_2} = 0.3408 \text{ mol } NaHCO_3$$

$$0.3408 \text{ mol } NaHCO_3 \times \frac{84.01 \text{ g } NaHCO_3}{1 \text{ mol } NaHCO_3} = 28.6 \text{ g } NaHCO_3$$

70. $C_6H_{12}O_6(s) + 6O_2(g) \rightarrow 6CO_2(g) + 6H_2O(g)$

molar masses: glucose, 180.2 g; CO_2, 44.01 g

116   Chapter 10   Chemical Quantities

$$1.00 \text{ g glucose} \times \frac{1 \text{ mol glucose}}{180.2 \text{ g glucose}} = 5.549 \times 10^{-3} \text{ mol glucose}$$

$$5.549 \times 10^{-3} \text{ mol glucose} \times \frac{6 \text{ mol CO}_2}{1 \text{ mol glucose}} = 3.32 \times 10^{-2} \text{ mol CO}_2$$

$$3.32 \times 10^{-2} \text{ mol CO}_2 \times \frac{44.01 \text{ g CO}_2}{1 \text{ mol CO}_2} = 1.47 \text{ g CO}_2$$

72. $Pb^{2+}(aq) + CrO_4^{2-}(aq) \rightarrow PbCrO_4(s)$

    millimolar ionic masses: Pb^{2+}, 207.2 mg; K_2CrO_4, 194.2 mg

    $$15 \text{ mg Pb}^{2+} \times \frac{1 \text{ mmol Pb}^{2+}}{207.2 \text{ mg Pb}^{2+}} = 0.07239 \text{ mmol Pb}^{2+}$$

    $$0.07239 \text{ mmol Pb}^{2+} \times \frac{1 \text{ mmol CrO}_4^{2-}}{1 \text{ mmol Pb}^{2+}} = 0.07239 \text{ mmol CrO}_4^{2-}$$

    $$0.07239 \text{ mmol CrO}_4^{2-} = 0.07239 \text{ mmol K}_2\text{CrO}_4$$

    $$0.07239 \text{ mmol CrO}_4^{2-} \times \frac{194.2 \text{ mg K}_2\text{CrO}_4}{1 \text{ mmol K}_2\text{CrO}_4} = 14 \text{ mg K}_2\text{CrO}_4$$

74. a. $UO_2(s) + 4HF(aq) \rightarrow UF_4(aq) + 2H_2O(l)$

    One molecule of uranium(IV) oxide will combine with four molecules of hydrofluoric acid, producing one uranium(IV) fluoride molecule and two water molecules. One mole of uranium(IV) oxide will combine with four moles of hydrofluoric acid to produce one mole of uranium(IV) fluoride and two moles of water.

    b. $2NaC_2H_3O_2(aq) + H_2SO_4(aq) \rightarrow Na_2SO_4(aq) + 2HC_2H_3O_2(aq)$

    Two molecules (formula units) of sodium acetate react exactly with one molecule of sulfuric acid, producing one molecule (formula unit) of sodium acetate and two molecules of acetic acid. Two moles of sodium acetate will combine with one mole of sulfuric acid, producing one mole of sodium sulfate and two moles of acetic acid.

    c. $Mg(s) + 2HCl(aq) \rightarrow MgCl_2(aq) + H_2(g)$

    One magnesium atom will react with two hydrochloric acid molecules (formula units) to produce one molecule (formula unit) of magnesium chloride and one molecule of hydrogen gas. One mole of magnesium will combine with two moles of hydrochloric acid, producing one mole of magnesium chloride and one mole of gaseous hydrogen.

    d. $B_2O_3(s) + 3H_2O(l) \rightarrow 2B(OH)_3(s)$

    One molecule (formula unit) of diboron trioxide will react exactly with three molecules of water, producing two molecules of boron trihydroxide (boric acid). One mole of diboron trioxide will combine

Chapter 10 Chemical Quantities 117

with three moles of water to produce two moles of boron trihydroxide (boric acid).

76. For O_2: $\dfrac{5 \text{ mol } O_2}{1 \text{ mol } C_3H_8}$ For CO_2: $\dfrac{3 \text{ mol } CO_2}{1 \text{ mol } C_3H_8}$ For H_2O: $\dfrac{4 \text{ mol } H_2O}{1 \text{ mol } C_3H_8}$

78. a. $NH_3(g) + HCl(g) \rightarrow NH_4Cl(s)$

   molar mass of NH_3 = 17.0 g

   $1.00 \text{ g } NH_3 \times \dfrac{1 \text{ mol } NH_3}{17.0 \text{ g } NH_3} = 0.0588 \text{ mol } NH_3$

   $0.0588 \text{ mol } NH_3 \times \dfrac{1 \text{ mol } NH_4Cl}{1 \text{ mol } NH_3} = 0.0588 \text{ mol } NH_4Cl$

   b. $CaO(s) + CO_2(g) \rightarrow CaCO_3(s)$

   molar mass CaO = 56.1 g

   $1.00 \text{ g CaO} \times \dfrac{1 \text{ mol CaO}}{56.1 \text{ g CaP}} = 0.0178 \text{ mol CaO}$

   $0.0178 \text{ mol CaO} \times \dfrac{1 \text{ mol } CaCO_3}{1 \text{ mol CaO}} = 0.0178 \text{ mol } CaCO_3$

   c. $4Na(s) + O_2(g) \rightarrow 2Na_2O(s)$

   molar mass Na = 23.0 g

   $1.00 \text{ g Na} \times \dfrac{1 \text{ mol Na}}{22.99 \text{ g Na}} = 0.0435 \text{ mol Na}$

   $0.0435 \text{ mol Na} \times \dfrac{2 \text{ mol } Na_2O}{4 \text{ mol Na}} = 0.0217 \text{ mol } Na_2O$

   d. $2P(s) + 3Cl_2(g) \rightarrow 2PCl_3(l)$

   molar mass P = 30.97 g

   $1.00 \text{ g P} \times \dfrac{1 \text{ mol P}}{30.97 \text{ g P}} = 0.0323 \text{ mol P}$

   $0.0323 \text{ mol P} \times \dfrac{2 \text{ mol } PCl_3}{2 \text{ mol P}} = 0.0323 \text{ mol } PCl_3$

80. a. molar mass HNO_3 = 63.0 g

   $5.0 \text{ mol } HNO_3 \times \dfrac{63.0 \text{ g } HNO_3}{1 \text{ mol } HNO_3} = 3.2 \times 10^2 \text{ g } HNO_3$

118   Chapter 10   Chemical Quantities

b.   molar mass Hg = 200.6 g

$$0.000305 \text{ mol Hg} \times \frac{200.6 \text{ g Hg}}{1 \text{ mol Hg}} = 0.0612 \text{ g Hg}$$

c.   molar mass K_2CrO_4 = 194.2 g

$$2.31 \times 10^{-5} \text{ mol } K_2CrO_4 \times \frac{194.2 \text{ g } K_2CrO_4}{1 \text{ mol } K_2CrO_4} = 4.49 \times 10^{-3} \text{ g } K_2CrO_4$$

d.   molar mass $AlCl_3$ = 133.3 g

$$10.5 \text{ mol } AlCl_3 \times \frac{133.3 \text{ g } AlCl_3}{1 \text{ mol } AlCl_3} = 1.40 \times 10^3 \text{ g } AlCl_3$$

e.   molar mass SF_6 = 146.1 g

$$4.9 \times 10^4 \text{ mol } SF_6 \times \frac{146.1 \text{ g } SF_6}{1 \text{ mol } SF_6} = 7.2 \times 10^6 \text{ g } SF_6$$

f.   molar mass NH_3 = 17.0 g

$$125 \text{ mol } NH_3 \times \frac{17.0 \text{ g } NH_3}{1 \text{ mol } NH_3} = 2.13 \times 10^3 \text{ g } NH_3$$

g.   molar mass Na_2O_2 = 77.98 g

$$0.01205 \text{ mol } Na_2O_2 \times \frac{77.98 \text{ g } Na_2O_2}{1 \text{ mol } Na_2O_2} = 0.9397 \text{ g } Na_2O_2$$

82. $2SO_2(g) + O_2(g) \rightarrow 2SO_3(g)$
molar masses: SO_2, 64.07 g; SO_3, 80.07 g
150 kg = 1.5×10^5 g

$$1.5 \times 10^5 \text{ g } SO_2 \times \frac{1 \text{ mol } SO_2}{64.07 \text{ g } SO_2} = 2.34 \times 10^3 \text{ mol } SO_2$$

$$2.34 \times 10^3 \text{ mol } SO_2 \times \frac{2 \text{ mol } SO_3}{2 \text{ mol } SO_2} = 2.34 \times 10^3 \text{ mol } SO_3$$

$$2.34 \times 10^3 \text{ mol } SO_3 \times \frac{80.07 \text{ g } SO_3}{1 \text{ mol } SO_3} = 1.9 \times 10^5 \text{ g } SO_3 = 1.9 \times 10^2 \text{ kg } SO_3$$

84. $Cl_2(g) + 2NaBr(aq) \rightarrow 2NaCl(aq) + Br_2(l)$
molar masses: Cl_2, 70.90 g; Br_2, 159.80 g

$$25.0 \text{ g } Cl_2 \times \frac{1 \text{ mol } Cl_2}{70.90 \text{ g } Cl_2} = 0.3526 \text{ mol } Cl_2$$

$$0.3526 \text{ mol } Cl_2 \times \frac{1 \text{ mol } Br_2}{1 \text{ mol } Cl_2} = 0.3526 \text{ mol } Br_2$$

$$0.3526 \text{ mol Br}_2 \times \frac{159.80 \text{ g Br}_2}{1 \text{ mol Br}_2} = 56.3 \text{ g Br}_2$$

86. $Zn(s) + 2HCl(aq) \rightarrow ZnCl_2(aq) + H_2(g)$

    molar masses: Zn, 65.38 g; H_2, 2.016 g

    $$2.50 \text{ g Zn} \times \frac{1 \text{ mol Zn}}{65.38 \text{ g Zn}} = 0.03824 \text{ mol Zn}$$

    $$0.03824 \text{ mol Zn} \times \frac{1 \text{ mol H}_2}{1 \text{ mol Zn}} = 0.03824 \text{ mol H}_2$$

    $$0.03824 \text{ mol H}_2 \times \frac{2.016 \text{ g H}_2}{1 \text{ mol H}_2} = 0.0771 \text{ g H}_2$$

88. a. $2Na(s) + Br_2(l) \rightarrow 2NaBr(s)$

    molar masses: Na, 22.99 g; Br_2, 159.8 g; NaBr, 102.9 g

    $$5.0 \text{ g Na} \times \frac{1 \text{ mol Na}}{22.99 \text{ g Na}} = 0.2175 \text{ mol Na}$$

    $$5.0 \text{ g Br}_2 \times \frac{1 \text{ mol Br}_2}{159.8 \text{ g Br}_2} = 0.03129 \text{ mol Br}_2$$

    Intuitively, we would suspect that Br_2 is the limiting reactant, since there is much less Br_2 than Na on a mole basis. To *prove* that Br_2 is the limiting reactant, the following calculation is needed:

    $$0.03129 \text{ mol Br}_2 \times \frac{2 \text{ mol Na}}{1 \text{ mol Br}_2} = 0.06258 \text{ mol Na}$$

    Clearly, there is more Na than this present, so Br_2 limits the reaction extent and the amount of NaBr formed.

    $$0.03129 \text{ mol Br}_2 \times \frac{2 \text{ mol NaBr}}{1 \text{ mol Br}_2} = 0.06258 \text{ mol NaBr}$$

    $$0.06258 \text{ mol NaBr} \times \frac{102.9 \text{ g NaBr}}{1 \text{ mol NaBr}} = 6.4 \text{ g NaBr}$$

    b. $Zn(s) + CuSO_4(aq) \rightarrow ZnSO_4(aq) + Cu(s)$

    molar masses: Zn, 65.38 g; Cu, 63.55 g; $ZnSO_4$, 161.5 g; $CuSO_4$, 159.6 g

    $$5.0 \text{ g Zn} \times \frac{1 \text{ mol Zn}}{65.38 \text{ g Zn}} = 0.07648 \text{ mol Zn}$$

    $$5.0 \text{ g CuSO}_4 \times \frac{1 \text{ mol CuSO}_4}{159.6 \text{ g CuSO}_4} = 0.03132 \text{ mol CuSO}_4$$

Since the coefficients of Zn and $CuSO_4$ are the *same* in the balanced chemical equation, an equal number of moles of Zn and $CuSO_4$ would be needed for complete reaction. Since there is less $CuSO_4$ present, $CuSO_4$ must clearly be the limiting reactant.

$$0.03132 \text{ mol } CuSO_4 \times \frac{1 \text{ mol } ZnSO_4}{1 \text{ mol } CuSO_4} = 0.03132 \text{ mol } ZnSO_4$$

$$0.03132 \text{ mol } ZnSO_4 \times \frac{161.5 \text{ g } ZnSO_4}{1 \text{ mol } ZnSO_4} = 5.1 \text{ g } ZnSO_4$$

$$0.03132 \text{ mol } CuSO_4 \times \frac{1 \text{ mol } Cu}{1 \text{ mol } CuSO_4} = 0.03132 \text{ mol } Cu$$

$$0.03132 \text{ mol } Cu \times \frac{63.55 \text{ g } Cu}{1 \text{ mol } Cu} = 2.0 \text{ g } Cu$$

c. $NH_4Cl(aq) + NaOH(aq) \rightarrow NH_3(g) + H_2O(l) + NaCl(aq)$

molar masses: NH_4Cl, 53.49 g; NaOH, 40.00 g; NH_3, 17.03 g

H_2O, 18.02 g; NaCl, 58.44 g

$$5.0 \text{ g } NH_4Cl \times \frac{1 \text{ mol } NH_4Cl}{53.49 \text{ g } NH_4Cl} = 0.09348 \text{ mol } NH_4Cl$$

$$5.0 \text{ g NaOH} \times \frac{1 \text{ mol NaOH}}{40.00 \text{ g NaOH}} = 0.1250 \text{ mol NaOH}$$

Since the coefficients of NH_4Cl and NaOH are both *one* in the balanced chemical equation for the reaction, an equal number of moles of NH_4Cl and NaOH would be needed for complete reaction. Since there is less NH_4Cl present, NH_4Cl must be the limiting reactant.

Since the coefficients of the products in the balanced chemical equation are also all *one*, if 0.09348 mol of NH_4Cl (the limiting reactant) reacts completely, then 0.09348 mol of each product will be formed.

$$0.09348 \text{ mol } NH_3 \times \frac{17.03 \text{ g } NH_3}{1 \text{ mol } NH_3} = 1.6 \text{ g } NH_3$$

$$0.09348 \text{ mol } H_2O \times \frac{18.02 \text{ g } H_2O}{1 \text{ mol } H_2O} = 1.7 \text{ g } H_2O$$

$$0.09348 \text{ mol NaCl} \times \frac{58.44 \text{ g NaCl}}{1 \text{ mol NaCl}} = 5.5 \text{ g NaCl}$$

d. $Fe_2O_3(s) + 3CO(g) \rightarrow 2Fe(s) + 3CO_2(g)$

molar masses: Fe_2O_3, 159.7 g; CO, 28.01 g

Fe, 55.85 g; CO_2, 44.01 g

$$5.0 \text{ g Fe}_2\text{O}_3 \times \frac{1 \text{ mol Fe}_2\text{O}_3}{159.7 \text{ g Fe}_2\text{O}_3} = 0.03131 \text{ mol Fe}_2\text{O}_3$$

$$5.0 \text{ g CO} \times \frac{1 \text{ mol CO}}{28.01 \text{ g CO}} = 0.1785 \text{ mol CO}$$

Because there is considerably less Fe_2O_3 than CO on a mole basis, let's see if Fe_2O_3 is the limiting reactant.

$$0.03131 \text{ mol Fe}_2\text{O}_3 \times \frac{3 \text{ mol CO}}{1 \text{ mol Fe}_2\text{O}_3} = 0.09393 \text{ mol CO}$$

Since there is 0.1785 mol of CO present, but we have determined that only 0.09393 mol CO would be needed to react with all the Fe_2O_3 present, then Fe_2O_3 must be the limiting reactant. CO is present in excess.

$$0.03131 \text{ mol Fe}_2\text{O}_3 \times \frac{2 \text{ mol Fe}}{1 \text{ mol Fe}_2\text{O}_3} = 0.06262 \text{ mol Fe}$$

$$0.06262 \text{ mol Fe} \times \frac{55.85 \text{ g Fe}}{1 \text{ mol Fe}} = 3.5 \text{ g Fe}$$

$$0.03131 \text{ mol Fe}_2\text{O}_3 \times \frac{3 \text{ mol CO}_2}{1 \text{ mol Fe}_2\text{O}_3} = 0.09393 \text{ mol CO}_2$$

$$0.09393 \text{ mol CO}_2 \times \frac{44.01 \text{ g CO}_2}{1 \text{ mol CO}_2} = 4.1 \text{ g CO}_2$$

90. $N_2H_4(l) + O_2(g) \rightarrow N_2(g) + 2H_2O(g)$

molar masses: N_2H_4, 32.05 g; O_2, 32.00 g; N_2, 28.02 g; H_2O, 18.02 g

$$20.0 \text{ g N}_2\text{H}_4 \times \frac{1 \text{ mol N}_2\text{H}_4}{32.05 \text{ g N}_2\text{H}_4} = 0.624 \text{ mol N}_2\text{H}_4$$

$$20.0 \text{ g O}_2 \times \frac{1 \text{ mol O}_2}{32.00 \text{ g O}_2} = 0.625 \text{ mol O}_2$$

The two reactants are present in nearly the required ratio for complete reaction (due to the 1:1 stoichiometry of the reaction and the very similar molar masses of the substances). We will consider N_2H_4 as the limiting reactant in the following calculations.

$$0.624 \text{ mol N}_2\text{H}_4 \times \frac{1 \text{ mol N}_2}{1 \text{ mol N}_2\text{H}_4} = 0.624 \text{ mol N}_2$$

$$0.624 \text{ mol N}_2 \times \frac{28.02 \text{ g N}_2}{1 \text{ mol N}_2} = 17.5 \text{ g N}_2$$

$$0.624 \text{ mol N}_2\text{H}_4 \times \frac{2 \text{ mol H}_2\text{O}}{1 \text{ mol N}_2\text{H}_4} = 1.248 \text{ mol H}_2\text{O} = 1.25 \text{ mol H}_2\text{O}$$

$$1.248 \text{ mol H}_2\text{O} \times \frac{18.02 \text{ g H}_2\text{O}}{1 \text{ mol H}_2\text{O}} = 22.5 \text{ g H}_2\text{O}$$

92. $$12.5 \text{ g theory} \times \frac{40 \text{ g actual}}{100 \text{ g theory}} = 5 \text{ g}$$

Cumulative Review: Chapters 9 and 10

2. On a microscopic basis, one mole of a substance represents Avogadro's number (6.022×10^{23}) of individual units (atoms or molecules) of the substance. On a macroscopic, more practical basis, one mole of a substance represents the amount of substance present when the molar mass of the substance in grams is taken (for example 12.01 g of carbon will be one mole of carbon). Chemists have chosen these definitions so that there will be a simple relationship between measurable amounts of substances (grams) and the actual number of atoms or molecules present, and so that the number of particles present in samples of *different* substances can easily be compared. For example, it is known that carbon and oxygen react by the reaction

$$C(s) + O_2(g) \rightarrow CO_2(g)$$

   Chemists understand this equation to mean that one carbon atom reacts with one oxygen molecule to produce one molecule of carbon dioxide, and also that one mole (12.01 g) of carbon will react with one mole (32.00 g) of oxygen to produce one mole (44.01 g) of carbon dioxide.

4. The molar mass of a compound is the mass in grams of one mole of the compound (6.022×10^{23} molecules of the compound), and is calculated by summing the average atomic masses of all the atoms present in a molecule of the compound. For example, a molecule of the compound H_3PO_4 contains three hydrogen atoms, one phosphorus atom, and four oxygen atom: the molar mass is obtained by adding up the average atomic masses of these atoms.

   molar mass H_3PO_4 = 3(1.008 g) + 1(30.97 g) + 4(16.00 g) = 97.99 g

6. The empirical formula of a compound represents just the relative number of atoms of each type present in a molecule of the compound, whereas the molecular formula represents the actual number of atoms of each type present in a real molecule of the compound. For example, both acetylene (molecular formula C_2H_2) and benzene (molecular formula C_6H_6) have the same relative number of carbon and hydrogen atoms (one hydrogen for each carbon atom), and so have the same empirical formula (CH). Once the empirical formula of a compound has been determined, it is also necessary to determine the molar mass of the compound before the actual molecular formula can be calculated. Since real molecules cannot contain fractional parts of atoms, the molecular formula is always a whole number multiple of the empirical formula. For the examples above, the molecular formula of acetylene is twice the empirical formula, and the molecular formula of benzene is six times the empirical formula (both factors are integers).

8. In question 7, suppose we had chosen to calculate the percentage composition of phosphoric acid, H_3PO_4: 3.086% H, 31.60% P, 65.31% O. We could convert this percentage composition data into "experimental" data by first choosing a mass of sample to be "analyzed", and then calculating what mass of each element is present in this size sample using the percentage of each element. For example, suppose we choose our sample to have a mass of 2.417 g. Then the masses of H, P, and O present in this sample would be given by the following

$$\text{g H} = (2.417 \text{ g sample}) \times \frac{3.086 \text{ g H}}{100.0 \text{ g sample}} = 0.07459 \text{ g H}$$

$$\text{g P} = (2.417 \text{ g sample}) \times \frac{31.60 \text{ g P}}{100.0 \text{ g sample}} = 0.7638 \text{ g P}$$

$$\text{g O} = (2.417 \text{ g sample}) \times \frac{65.31 \text{ g O}}{100.0 \text{ g sample}} = 1.579 \text{ g O}$$

Note that (0.07459 g + 0.7638 g + 1.579 g) = 2.41739 = 2.417 g.

So our new problem could be worded as follows: "A 2.417-g sample of a compound has been analyzed and was found to contain 0.07459 g H, 0.7638 g of P, and 1.579 g of oxygen. Calculate the empirical formula of the compound".

$$\text{mol H} = (0.07459 \text{ g H}) \times \frac{1 \text{ mol H}}{1.008 \text{ g H}} = 0.07400 \text{ mol H}$$

$$\text{mol P} = (0.7638 \text{ g P}) \times \frac{1 \text{ mol P}}{30.97 \text{ g P}} = 0.02466 \text{ mol P}$$

$$\text{mol O} = (1.579 \text{ g O}) \times \frac{1 \text{ mol O}}{16.00 \text{ g O}} = 0.09869 \text{ mol O}$$

Dividing each of these numbers of moles by the smallest number of moles (0.02466 mol P) gives the following

$$\frac{0.07400 \text{ mol H}}{0.02466} = 3.001 \text{ mol H}$$

$$\frac{0.02466 \text{ mol P}}{0.02466} = 1.000 \text{ mol P}$$

$$\frac{0.09869 \text{ mol O}}{0.02466} = 4.002 \text{ mol O}$$

and the empirical formula is (not surprisingly) just H_3PO_4!

10. The mole ratios for a reaction are based on the *coefficients* of the balanced chemical equation for the reaction: these coefficients show in what proportions molecules (or moles of molecules) combine. For a given amount of C_4H_{10}, the following mole ratios could be constructed, which would enable you to calculate the number of moles of each product, or of the second reactant, that would be involved.

$$2C_4H_{10}(g) + 13O_2(g) \rightarrow 8CO_2(g) + 10H_2O(g)$$

to calculate mol CO_2 produced: $\dfrac{8 \text{ mol } CO_2}{2 \text{ mol } C_4H_{10}}$

to calculate mol H_2O produced: $\dfrac{10 \text{ mol } H_2O}{2 \text{ mol } C_4H_{10}}$

to calculate mol O_2 required: $\dfrac{13 \text{ mol } O_2}{2 \text{ mol } C_4H_{10}}$

We could then calculate the numbers of moles of the other substances if 0.55 mol of C_4H_{10} were to be combusted as follows:

mol CO_2 produced = $(0.55 \text{ mol } C_4H_{10}) \times \dfrac{8 \text{ mol } CO_2}{2 \text{ mol } C_4H_{10}}$ = 2.2 mol CO_2

mol H_2O produced = $(0.55 \text{ mol } C_4H_{10}) \times \dfrac{10 \text{ mol } H_2O}{2 \text{ mol } C_4H_{10}}$ = 2.8 mol H_2O

mol O_2 required = $(0.55 \text{ mol } C_4H_{10}) \times \dfrac{13 \text{ mol } O_2}{2 \text{ mol } C_4H_{10}}$ = 3.6 mol O_2

12. Although we can calculate specifically the exact amounts of each reactant needed for a chemical reaction, oftentimes reaction mixtures are prepared using more or less arbitrary amounts of the reagents. However, regardless of how much of each reagent may be taken for a reaction, the substances still react stoichiometrically, according to the mole ratios derived from the balanced chemical equation for the reaction. When arbitrary amounts of reactants are used, there will be one reactant which, stoichiometrically, is present in the least amount. This substance is called the *limiting reactant* for the experiment. It is the limiting reactant that controls how much product is formed, regardless of how much of the other reactants are present. The limiting reactant limits the amount of product that can form in the experiment, because once the limiting reactant has reacted completely, the reaction must stop. We say that the other reactants in the experiment are present in excess, which means that a portion of these reactants will still be present unchanged after the reaction has ended and the limiting reactant has been used up completely.

14. The *theoretical yield* for an experiment is the mass of product calculated based on the limiting reactant for the experiment being completely consumed. The *actual yield* for an experiment is the mass of product actually collected by the experimenter. Obviously, any experiment is restricted by the skills of the experimenter and by the inherent limitations of the experimental method being used. For these reasons, the actual yield is often *less* than the theoretical yield (most scientific writers report the actual or percentage yield for their experiments as an indication of the usefulness of their experiments). Although one would expect that the actual yield should never be more than the theoretical yield, in real experiments, sometimes this happens: however, an actual yield greater than a theoretical yield is usually taken to mean that something is *wrong* in either the experiment (for example, impurities may be present, or the reaction may not occur as envisioned) or in the calculations.

16. a. $\%H = \dfrac{2(1.008 \text{ g})}{18.02 \text{ g}} \times 100 = 11.19\%H$

   $\%O = \dfrac{16.00 \text{ g}}{18.02 \text{ g}} \times 100 = 88.81\%O$

   b. $\%K = \dfrac{2(39.10 \text{ g})}{138.2 \text{ g}} \times 100 = 56.58\%K$

   $\%C = \dfrac{12.01 \text{ g}}{138.2 \text{ g}} \times 100 = 8.690\%C$

   $\%O = \dfrac{3(16.00 \text{ g})}{138.2 \text{ g}} \times 100 = 34.73\%O$

   c. $\%N = \dfrac{14.01 \text{ g}}{46.01 \text{ g}} \times 100 = 30.45\%N$

   $\%O = \dfrac{2(16.00 \text{ g})}{46.01 \text{ g}} \times 100 = 69.55\%O$

   d. $\%C = \dfrac{12(12.01 \text{ g})}{342.3 \text{ g}} \times 100 = 42.10\%C$

   $\%H = \dfrac{22(1.008 \text{ g})}{342.3 \text{ g}} \times 100 = 6.479\%H$

   $\%O = \dfrac{11(16.00 \text{ g})}{342.3 \text{ g}} \times 100 = 51.42\%O$

   e. $\%Na = \dfrac{22.99 \text{ g}}{40.00 \text{ g}} \times 100 = 57.48\%Na$

   $\%O = \dfrac{16.00 \text{ g}}{40.00 \text{ g}} \times 100 = 40.00\%O$

   $\%H = \dfrac{1.008 \text{ g}}{40.00 \text{ g}} \times 100 = 2.520\%H$

   f. $\%Na = \dfrac{2(22.99 \text{ g})}{77.98 \text{ g}} \times 100 = 58.96\%Na$

   $\%O = \dfrac{2(16.00 \text{ g})}{77.98 \text{ g}} \times 100 = 41.04\%O$

   g. $\%Na = \dfrac{22.99 \text{ g}}{82.03 \text{ g}} \times 100 = 28.03\%Na$

   $\%C = \dfrac{2(12.01 \text{ g})}{82.03 \text{ g}} \times 100 = 29.28\%C$

$$\%H = \frac{3(1.008 \text{ g})}{82.03 \text{ g}} \times 100 = 3.686\%H$$

$$\%O = \frac{2(16.00 \text{ g})}{82.03 \text{ g}} \times 100 = 39.01\%O$$

18. a. $2AgNO_3(aq) + CaSO_4(aq) \rightarrow Ag_2SO_4(s) + Ca(NO_3)_2(aq)$

   Molar masses: $AgNO_3$, 169.9 g; Ag_2SO_4, 311.9 g; $Ca(NO_3)_2$, 164.1 g

   $$25.0 \text{ g AgNO}_3 \times \frac{1 \text{ mol}}{169.9 \text{ g}} = 0.147 \text{ mol AgNO}_3$$

   $$0.147 \text{ mol AgNO}_3 \times \frac{1 \text{ mol Ag}_2\text{SO}_4}{2 \text{ mol AgNO}_3} = 0.0735 \text{ mol Ag}_2\text{SO}_4$$

   $$0.0735 \text{ mol Ag}_2\text{SO}_4 \times \frac{311.9 \text{ g}}{1 \text{ mol}} = 22.9 \text{ g Ag}_2\text{SO}_4$$

   $$0.147 \text{ mol AgNO}_3 \times \frac{1 \text{ mol Ca(NO}_3)_2}{2 \text{ mol AgNO}_3} = 0.0735 \text{ mol Ca(NO}_3)_2$$

   $$0.0735 \text{ mol Ca(NO}_3)_2 \times \frac{164.1 \text{ g}}{1 \text{ mol}} = 12.1 \text{ g Ca(NO}_3)_2$$

b. $2Al(s) + 6HNO_3(aq) \rightarrow 2Al(NO_3)_3(aq) + 3H_2(g)$

   Molar masses: Al, 26.98 g; $Al(NO_3)_3$, 213.0 g; H_2, 2.016 g

   $$25.0 \text{ g Al} \times \frac{1 \text{ mol}}{26.98 \text{ g}} = 0.927 \text{ mol Al}$$

   $$0.927 \text{ mol Al} \times \frac{2 \text{ mol Al(NO}_3)_3}{2 \text{ mol Al}} = 0.927 \text{ mol Al(NO}_3)_3$$

   $$0.927 \text{ mol Al(NO}_3)_3 \times \frac{213.0 \text{ g}}{1 \text{ mol}} = 197 \text{ g Al(NO}_3)_3$$

   $$0.927 \text{ mol Al} \times \frac{3 \text{ mol H}_2}{2 \text{ mol Al}} = 1.39 \text{ mol H}_2$$

   $$1.39 \text{ mol H}_2 \times \frac{2.016 \text{ g}}{1 \text{ mol}} = 2.80 \text{ g H}_2$$

c. $H_3PO_4(aq) + 3NaOH(aq) \rightarrow Na_3PO_4(aq) + 3H_2O(l)$

   Molar masses: H_3PO_4, 97.99 g; Na_3PO_4, 163.9 g; H_2O, 18.02 g

   $$25.0 \text{ g H}_3\text{PO}_4 \times \frac{1 \text{ mol}}{97.99 \text{ g}} = 0.255 \text{ mol H}_3\text{PO}_4$$

$$0.255 \text{ mol } H_3PO_4 \times \frac{1 \text{ mol } Na_3PO_4}{1 \text{ mol } H_3PO_4} = 0.255 \text{ mol } Na_3PO_4$$

$$0.255 \text{ mol } Na_3PO_4 \times \frac{163.9 \text{ g}}{1 \text{ mol}} = 41.8 \text{ g } Na_3PO_4$$

$$0.255 \text{ mol } H_3PO_4 \times \frac{3 \text{ mol } H_2O}{1 \text{ mol } H_3PO_4} = 0.765 \text{ mol } H_2O$$

$$0.765 \text{ mol } H_2O \times \frac{18.02 \text{ g}}{1 \text{ mol}} = 13.8 \text{ g } H_2O$$

d. $CaO(s) + 2HCl(aq) \rightarrow CaCl_2(aq) + H_2O(l)$

Molar masses: CaO, 56.08 g; $CaCl_2$, 111.0 g; H_2O, 18.02 g

$$25.0 \text{ g CaO} \times \frac{1 \text{ mol}}{56.08 \text{ g}} = 0.446 \text{ mol CaO}$$

$$0.446 \text{ mol CaO} \times \frac{1 \text{ mol } CaCl_2}{1 \text{ mol CaO}} = 0.446 \text{ mol } CaCl_2$$

$$0.446 \text{ mol } CaCl_2 \times \frac{111.0 \text{ g}}{1 \text{ mol}} = 49.5 \text{ g } CaCl_2$$

$$0.446 \text{ mol CaO} \times \frac{1 \text{ mol } H_2O}{1 \text{ mol CaO}} = 0.446 \text{ mol } H_2O$$

$$0.446 \text{ mol } H_2O \times \frac{18.02 \text{ g}}{1 \text{ mol}} = 8.04 \text{ g } H_2O$$

20. For sodium: $2Na(s) + Cl_2(g) \rightarrow 2NaCl(s)$

Molar masses: Na, 22.99 g; Cl_2, 70.90 g; NaCl, 58.44 g

$$10.0 \text{ g Na} \times \frac{1 \text{ mol}}{22.99 \text{ g}} = 0.435 \text{ mol Na}$$

$$10.0 \text{ g } Cl_2 \times \frac{1 \text{ mol}}{70.90 \text{ g}} = 0.141 \text{ mol } Cl_2$$

Cl_2 is the limiting reactant

$$0.141 \text{ mol } Cl_2 \times \frac{2 \text{ mol NaCl}}{1 \text{ mol } Cl_2} \times \frac{58.44 \text{ g}}{1 \text{ mol}} = 16.5 \text{ g NaCl}$$

$$\% \text{ yield} = \frac{9.85 \text{ g actual}}{16.5 \text{ g theory}} \times 100 = 59.7\%$$

For magnesium: $Mg(s) + Cl_2(g) \rightarrow MgCl_2(s)$

Molar masses: Mg, 24.31 g; Cl_2, 70.90 g; $MgCl_2$, 95.21 g

$$10.0 \text{ g Mg} \times \frac{1 \text{ mol}}{24.31 \text{ g}} = 0.411 \text{ mol Mg}$$

$$10.0 \text{ g Cl}_2 \times \frac{1 \text{ mol}}{70.90 \text{ g}} = 0.141 \text{ mol Cl}_2$$

Cl_2 is the limiting reactant

$$0.141 \text{ mol Cl}_2 \times \frac{1 \text{ mol MgCl}_2}{1 \text{ mol Cl}_2} \times \frac{95.21 \text{ g}}{1 \text{ mol}} = 13.4 \text{ g MgCl}_2$$

$$\% \text{ yield} = \frac{9.85 \text{ g actual}}{13.4 \text{ g theory}} \times 100 = 73.5\%$$

For aluminum: $2Al(s) + 3Cl_2(g) \rightarrow 2AlCl_3(s)$

Molar masses: Al, 26.98 g; Cl_2 70.90 g; $AlCl_3$, 133.3 g

$$10.0 \text{ g Al} \times \frac{1 \text{ mol}}{26.98 \text{ g}} = 0.371 \text{ mol Al}$$

$$10.0 \text{ g Cl}_2 \times \frac{1 \text{ mol}}{70.90 \text{ g}} = 0.141 \text{ mol Cl}_2$$

Cl_2 is the limiting reactant

$$0.141 \text{ mol Cl}_2 \times \frac{2 \text{ mol AlCl}_3}{3 \text{ mol Cl}_2} \times \frac{133.3 \text{ g}}{1 \text{ mol}} = 12.5 \text{ g AlCl}_3$$

$$\% \text{ yield} = \frac{9.85 \text{ g actual}}{12.5 \text{ g theory}} \times 100 = 78.8\%$$

Chapter 11   Modern Atomic Theory

2. The different forms of electromagnetic radiation are similar in that they all exhibit the same type of wave-like behavior and are propagated through space at the same speed (the speed of light). The types of electromagnetic radiation differ in their frequency (and wavelength) and in the resulting amount of energy carried per photon.

4. The *frequency* of electromagnetic radiation represents how many complete cycles of the wave pass a given point per second. The *speed* of electromagnetic radiation represents how fast a given wave moves through space.

6. Although electromagnetic radiation exhibits characteristic wave-like properties, such radiation also demonstrates a particle-like nature. For example, when an excited atom emits radiation, the radiation demonstrates wave properties, but occurs in discrete particle-like bundles (photons). The wave-particle nature of light refers to the fact that a beam of electromagnetic energy can be considered not only as a continuous wave, but also as a stream of discrete packets of energy moving through space.

8. An atom in an excited state returns to its ground state by emitting a photon having an energy corresponding to the energy difference between the two states.

10. larger ($E = h\nu$)

12. When excited hydrogen atoms emit their excess energy, the photons of radiation emitted are always of exactly the same wavelength and energy. We consider this to mean that the hydrogen atom possesses only certain allowed energy states, and that the photons emitted correspond to the atom changing from one of these allowed energy states to another of the allowed energy state. The energy of the photon emitted corresponds to the energy difference in the allowed states. If the hydrogen atom did not possess discrete energy levels, then we would expect the photons emitted to have random wavelengths and energies.

14. lower energy state [often the ground (lowest energy) state]

16. The ground state of an atom is its lowest possible energy state.

18. According to Bohr, electrons move in discrete, fixed circular *orbits* around the nucleus. If the wavelength of the applied energy corresponds to the *difference in energy* between the two orbits, the atom absorbs a photon and the electron moves to a larger orbit.

20. Bohr's theory *explained* the experimentally *observed* line spectrum of hydrogen *exactly*. Bohr's theory was ultimately discarded because when attempts were made to extend the theory to atoms other than hydrogen, the calculated properties did *not* correspond closely to experimental measurements.

22. An orbit represents a definite, exact circular pathway around the nucleus in which an electron can be found. An orbital represents a region of space in which there is a high probability of finding the electron.

24. Any experiment which sought to measure the exact location of an electron (such as shooting a beam of light at it) would cause the electron to move. Any measurement made would necessitate the application or removal of energy, which would disturb the electron from where it had been before the measurement.

26. Pictures we draw to represent orbitals should only be interpreted as probability maps. They are not meant to represent that the electron moves only on the surface of, or within, the region drawn in the picture. Since the mathematical probability of finding the electron never actually becomes zero on moving outward from the nucleus, scientists have decided that pictures of orbitals should represent a 90% probability that the electron will be found inside the region depicted in the drawing (for 100% probability, the orbital would have to encompass all space).

28. The $2p$ orbitals have two lobes and are sometimes described as having a "dumbbell" shape. The individual $2p$ orbitals ($2p_x$, $2p_y$, $2p_z$) are alike in shape and energy; they differ only in the direction in which the lobes of the orbital are oriented.

30. increases; the energy of an electron increases with its principal quantum number.

32. The fourth principal energy level is divided into four sublevels: these sublevels are given the designations $4s$, $4p$, $4d$, and $4f$.

34. The Pauli exclusion principle states that an orbital can hold a maximum of two electrons, and those two electrons must have opposite spins.

36. increases; as you move out from the nucleus, there is more space and room for more sublevels.

38. opposite

40. a. incorrect (the $n = 1$ shell has only the $1s$ subshell)
    b. incorrect (the $n = 2$ shell has only $2s$ and $2p$ subshells)
    c. correct (the $n = 4$ shell contains s, p, d and f subshells)
    d. correct (the $n = 5$ shell contains s, p, d, f, and g subshells)

42. When a hydrogen atom is in its ground state, the electron is found in the $1s$ orbital. The $1s$ orbital has the lowest energy of all the possible hydrogen orbitals.

132 Chapter 11 Modern Atomic Theory

44. The elements in a given vertical column of the periodic table have the same valence electron configuration.

46. a. $1s^2\ 2s^2\ 2p^6\ 3s^2\ 3p^6\ 4s^2\ 3d^{10}\ 4p^6\ 5s^2$
     b. $1s^2\ 2s^2\ 2p^6\ 3s^2\ 3p^6\ 4s^2\ 3d^{10}$
     c. $1s^2$
     d. $1s^2\ 2s^2\ 2p^6\ 3s^2\ 3p^6\ 4s^2\ 3d^{10}\ 4p^5$

48. a. $1s^2\ 2s^2\ 2p^2$
     b. $1s^2\ 2s^2\ 2p^6\ 3s^2\ 3p^3$
     c. $1s^2\ 2s^2\ 2p^6\ 3s^2\ 3p^4$
     d. $1s^2\ 2s^2\ 2p^1$

50. a. (↑↓) (↑↓) (↑↓)(↑↓)(↑↓) (↑↓) (↑)( )( )
          1s   2s   2p   3s   3p

     b. (↑↓) (↑↓) (↑↓)(↑↓)(↑↓) (↑↓) (↑)(↑)(↑)
          1s   2s   2p   3s   3p

     c. (↑↓) (↑↓) (↑↓)(↑↓)(↑↓) (↑↓) (↑↓)(↑↓)(↑↓) (↑↓)
          1s   2s   2p   3s   3p   4s
         (↑↓)(↑↓)(↑↓)(↑↓)(↑↓) (↑↓)(↑↓)(↑)
                   3d   4p

     d. (↑↓) (↑↓) (↑↓)(↑↓)(↑↓) (↑↓) (↑↓)(↑↓)(↑↓)
          1s   2s   2p   3s   3p

52. For the representative elements (those filling s and p subshells), the group number gives the number of valence electrons.

     a. two   (3s)
     b. four   (2s, 2p)
     c. seven (4s, 4p)
     d. one   (6s)

54. The properties of Rb and Sr suggest that they are members of Groups 1 and 2, respectively, and so must be filling the 5s orbital. The 5s orbital is lower in energy (and fills before) the 4d orbitals.

56. a. [Ar] $4s^2$
    b. [Rn] $7s^1$
    c. [Kr] $5s^2\ 4d^1$
    d. [Xe] $6s^2\ 4f^1\ 5d^1$

58. a. [Ne] $3s^2\ 3p^2$
    b. [Ar] $4s^2\ 3d^{10}\ 4p^1$
    c. [Ar] $4s^2\ 3d^3$
    d. [Ar] $4s^2\ 3d^1$

60. The number of 3d electrons can be found simply by *counting* along the first transition series (Z = 21 to Z = 30); e.g., the *third* element has *three* 3d electrons.

    a. one
    b. two
    c. zero
    d. ten

62. Figure 11.27 shows the orbitals being filled as a function of location in the periodic table.

    a. $5d$
    b. $4f$
    c. $5d$
    d. $3d$

64. Figure 11.30 shows partial electronic configurations.

    a. [Rn] $7s^2\ 5f^3\ 6d^1$
    b. [Ar] $4s^2\ 3d^5$
    c. [Xe] $6s^2\ 5d^{10}\ 4f^{14}$
    d. [Rn] $7s^1$

66. The metallic elements *lose* electrons and form *positive* ions (cations); the nonmetallic elements *gain* electrons and form *negative* ions (anions). Remember that the electron itself is *negatively* charged.

68. All exist as *diatomic* molecules (F_2, Cl_2, Br_2, I_2); all are *nonmetals*; all have relatively high electronegativities; all form 1− ions in reacting with metallic elements.

70. Elements at the *left* of a period (horizontal row) lose electrons more readily; at the left of a period (given principal energy level) the nuclear charge is the smallest and the electrons are least tightly held.

72. The elements of a given period (horizontal row) have valence electrons in the same subshells. Nuclear charge, however, increases across a period going from left to right. Atoms at the left side have smaller nuclear charges, and hold onto their valence electrons less tightly.

74. The *nuclear charge* increases from left to right within a period, pulling progressively more tightly on the valence electrons.

76. Ionization energies decrease in going from top to bottom within a vertical group; ionization energies increase in going from left to right within a horizontal period.

    a. Cs
    b. Li
    c. I
    d. Rb

78. Atomic size increases in going from top to bottom within a vertical group; atomic size decreases in going from left to right within a horizontal period.

    a. Na
    b. S
    c. N
    d. F

80. speed of light

82. photons

84. quantized

86. orbital

88. transition metal

90. spins (different m_s quantum numbers)

92. a. $1s^2\ 2s^2\ 2p^6\ 3s^2\ 3p^6\ 4s^1$ [Ar] $4s^1$

    (↑↓) (↑↓) (↑↓)(↑↓)(↑↓) (↑↓) (↑↓)(↑↓)(↑↓) (↑)
     1s   2s   2p   3s   3p   4s

b. $1s^2\ 2s^2\ 2p^6\ 3s^2\ 3p^6\ 4s^2\ 3d^2$ [Ar] $4s^2\ 3d^2$

   (↑↓) (↑↓) (↑↓)(↑↓)(↑↓) (↑↓) (↑↓)(↑↓)(↑↓) (↑↓) (↑)(↑)( )( )( )
   1s   2s   2p   3s   3p   4s   3d

c. $1s^2\ 2s^2\ 2p^6\ 3s^2\ 3p^2$ [Ne] $3s^2\ 3p^2$

   (↑↓) (↑↓) (↑↓)(↑↓)(↑↓) (↑↓) (↑)(↑)( )
   1s   2s   2p   3s   3p

d. $1s^2\ 2s^2\ 2p^6\ 3s^2\ 3p^6\ 4s^2\ 3d^6$ [Ar] $4s^2\ 3d^6$

   (↑↓) (↑↓) (↑↓)(↑↓)(↑↓) (↑↓) (↑↓)(↑↓)(↑↓) (↑↓) (↑↓)(↑)(↑)(↑)(↑)
   1s   2s   2p   3s   3p   4s   3d

e. $1s^2\ 2s^2\ 2p^6\ 3s^2\ 3p^6\ 4s^2\ 3d^{10}$ [Ar] $4s^2\ 3d^{10}$

   (↑↓) (↑↓) (↑↓)(↑↓)(↑↓) (↑↓) (↑↓)(↑↓)(↑↓) (↑↓) (↑↓)(↑↓)(↑↓)(↑↓)(↑↓)
   1s   2s   2p   3s   3p   4s   3d

94. a. ns^2 b. $ns^2\ np^5$
    c. $ns^2\ np^4$ d. ns^1
    e. $ns^2\ np^4$

96. a. $\lambda = \dfrac{h}{mv}$

   $\lambda = \dfrac{6.63 \times 10^{-34}\ \text{J s}}{(9.11 \times 10^{-31}\ \text{kg})[0.90(3.00 \times 10^8\ \text{m s}^{-1})]}$

   $\lambda = 2.7 \times 10^{-12}$ m (0.0027 nm)

   b. 4.4×10^{-34} m

   c. 2×10^{-35} m

   The wavelengths for the ball and the person are *infinitesimally small*, whereas the wavelength for the electron is nearly the same order of magnitude as the diameter of a typical atom.

98. Light is emitted from the hydrogen atom only at certain fixed wavelengths. If the energy levels of hydrogen were *continuous*, a hydrogen atom would emit energy at all possible wavelengths.

100. The third principal energy level of hydrogen is divided into *three* sublevels (3s, 3p, and 3d); there is a *single* 3s orbital; there is a set of *three* 3p orbitals; there is a set of *five* 3d orbitals.

136   Chapter 11   Modern Atomic Theory

102. a. incorrect; the $n = 1$ energy shell has only the $1s$ subshell
b. correct
c. incorrect; the $n = 3$ energy shell has only $3s$. $3p$, and $3d$ subshells
d. correct
e. correct
f. correct

104. a. $1s^2\ 2s^2\ 2p^6\ 3s^2\ 3p^5$ b. $1s^2\ 2s^2\ 2p^4$
c. $1s^2\ 2s^2\ 2p^6\ 3s^2\ 3p^1$ d. $1s^2\ 2s^1$

106. a. five ($2s$, $2p$)
b. seven ($3s$, $3p$)
c. one ($3s$)
d. three ($3s$, $3p$)

108. a. [Kr] $5s^2\ 4d^2$ b. [Kr] $5s^2\ 4d^{10}\ 5p^5$
c. [Ar] $4s^2\ 3d^{10}\ 4p^2$ d. [Xe] $6s^1$

110. The *position* of the element (both in terms of the vertical column and the horizontal row) tells you which set of orbitals is being filled last. See Figure 11.27 for details.

a. $3d$ b. $4d$
c. $5f$ d. $4p$

112. metals, low; nonmetals, high

114. Atomic size increases in going from top to bottom within a vertical group; atomic size decreases in going from left to right within a horizontal period.

a. Ba
b. Mg
c. Be

Chapter 12   Chemical Bonding

2. The strength of a chemical bond is characterized by the bond energy (the amount of energy required to break the bond).

4. ionic

6. The HF molecule contains atoms of two different elements. In bonding with each other, these atoms share a pair of valence electrons, but they do not share them equally. The bonding in HF is described as polar covalent, which indicates that although valence electron pairs are shared between atoms, the electron pair is attracted more strongly by one of the atoms than the other. Nonetheless, an HF molecule is more stable in energy than are separated H and F atoms.

8. polar

10. electronegativity

12. In each case, the element higher up within a given group of the periodic table, or to the right within a period, has the higher electronegativity.

   a.   K < Ca < Sc
   b.   At < Br < F
   c.   C < N < O

14. Generally, covalent bonds between atoms of *different* elements are *polar*.

   a.   covalent
   b.   polar covalent
   c.   polar covalent
   d.   ionic

16. For a bond to be polar covalent, the atoms involved in the bond must have different electronegativities (must be of different elements).

   a.   covalent (atoms of the same element)
   b.   covalent (atoms of the same element)
   c.   polar covalent (different elements)
   d.   polar covalent (different elements)

18. The *degree* of polarity of a polar covalent bond is indicated by the magnitude of the difference in electronegativities of the elements involved: the larger the difference in electronegativity, the more polar is the bond. Electronegativity differences are given in parentheses below:

   a.   H–O (1.4); H–N (0.9); the H–O bond is more polar.
   b.   H–N (0.9); H–F (1.9); the H–F bond is more polar.

137

c. H–O (1.4); H–F (1.9); the H–F bond is more polar.

d. H–O (1.4); H–Cl (0.9); the H–O bond is more polar.

20. The greater the electronegativity difference between two atoms, the more ionic will be the bond between those two atoms. Electronegativity differences are given in parentheses.

   a. Na–O (2.6) has more ionic character than Na–N (2.1)

   b. K–S (1.7) has more ionic character than K–P (1.3)

   c. K–Cl (2.2) has more ionic character than Na–Cl (2.1)

   d. Na–Cl (2.1) has more ionic character than Mg–Cl (1.8)

22. The presence of strong bond dipoles and a large overall dipole moment in water make it a very polar substance overall. Among those properties of water that are dependent on its dipole moment are its freezing point, melting point, vapor pressure, and its ability to dissolve many substances.

24. In a diatomic molecule containing two different elements, the more electronegative atom will be the negative end of the molecule, and the *less* electronegative atom will be the positive end.

   a. H

   b. Cl

   c. I

26. In the figures, the arrow points toward the more electronegative atom.

   a. $\delta+$ Cl→F $\delta-$

   b. $\delta+$ I→Br $\delta-$

   c. $\delta+$ Br→Cl $\delta-$

   d. $\delta+$ Br→F $\delta-$

28. In the figures, the arrow points toward the more electronegative atom.

   a. $\delta+$ P→S $\delta-$

   b. $\delta+$ S→O $\delta-$

   c. $\delta+$ S→N $\delta-$

   d. $\delta+$ S→Cl $\delta-$

30. previous

32. Atoms in covalent molecules gain a configuration like that of a noble gas by sharing one or more pairs of electrons between atoms: such shared

pairs of electrons "belong" to each of the atoms of the bond at the same time. In ionic bonding, one atom completely gives over one or more electrons to another atom, and then the resulting ions behave independently of one another.

34. a. Rb $1s^2\ 2s^2\ 2p^6\ 3s^2\ 3p^6\ 4s^2\ 3d^{10}\ 4p^6\ 5s^1$
    Rb$^+$ $1s^2\ 2s^2\ 2p^6\ 3s^2\ 3p^6\ 4s^2\ 3d^{10}\ 4p^6$
    Kr has the same configuration as Rb$^+$

    b. O $1s^2\ 2s^2\ 2p^4$
    O^{2-} $1s^2\ 2s^2\ 2p^6$
    Ne has the same configuration as O^{2-}

    c. Sr $1s^2\ 2s^2\ 2p^6\ 3s^2\ 3p^6\ 4s^2\ 3d^{10}\ 4p^6\ 5s^2$
    Sr^{2+} $1s^2\ 2s^2\ 2p^6\ 3s^2\ 3p^6\ 4s^2\ 3d^{10}\ 4p^6$
    Kr has the same configuration as Sr^{2+}

    d. Cs $1s^2\ 2s^2\ 2p^6\ 3s^2\ 3p^6\ 4s^2\ 3d^{10}\ 4p^6\ 5s^2\ 4d^{10}\ 5p^6\ 6s^1$
    Cs$^+$ $1s^2\ 2s^2\ 2p^6\ 3s^2\ 3p^6\ 4s^2\ 3d^{10}\ 4p^6\ 5s^2\ 4d^{10}\ 5p^6$
    Xe has the same configuration as Cs$^+$

    e. Ba $1s^2\ 2s^2\ 2p^6\ 3s^2\ 3p^6\ 4s^2\ 3d^{10}\ 4p^6\ 5s^2\ 4d^{10}\ 5p^6\ 6s^2$
    Ba^{2+} $1s^2\ 2s^2\ 2p^6\ 3s^2\ 3p^6\ 4s^2\ 3d^{10}\ 4p^6\ 5s^2\ 4d^{10}\ 5p^6$
    Xe has the same configuration as Ba^{2+}

36. a. Ca^{2+} (Ca has two electrons more than the noble gas Ar)
    b. N^{3-} (N has three electrons fewer than the noble gas Ne)
    c. Br$^-$ (Br has one electron fewer than the noble gas Kr)
    d. Sc$^+$ (Sc has one electron more than a filled 4s subshell)

38. a. CaO   Ca has two electrons more than a noble gas; O has two electrons fewer than a noble gas.

    b. SrF$_2$ Sr has two electrons more than a noble gas; F has one electron less than a noble gas.

    c. Al$_2$S$_3$ Al has three electrons more than a noble gas; S has two electrons fewer than a noble gas.

    d. Rb$_3$N Rb has one electron more than a noble gas; N has three electrons fewer than a noble gas.

e. Na$_2$Te   Na has one electron more than a noble gas; Te has two electrons fewer than a noble gas.

40. a. Al^{3+}, [Ne]; S^{2-}, [Ar]
    b. Mg^{2+}, [Ne]; N^{3-}, [Ne]
    c. Rb^{+}, [Kr]; O^{2-}, [Ne]
    d. Cs^{+}, [Xe]; I^{-}, [Xe]

42. An ionic solid such as NaCl basically consists of an array of alternating positively and negatively charged ions: that is, each positive ion has as its nearest neighbors a group of negative ions, and each negative ion has a group of positive ions surrounding it. In most ionic solids, the ions are packed as tightly as possible.

44. In forming an anion, an atom gains additional electrons in its outermost (valence) shell. Having additional electrons in the valence shell increases the repulsive forces between electrons, and the outermost shell becomes larger to accommodate this.

46. Relative ionic sizes are indicated in Figure 12.9. Within a given horizontal row of the periodic chart, negative ions tend to be larger than positive ions because the negative ions contain a larger number of electrons in the valence shell. Within a vertical group of the periodic table, ionic size increases from top to bottom. In general, positive ions are smaller than the atoms they come from, whereas negative ions are larger than the atoms they come from.

   a. F^{-}
   b. Cl^{-}
   c. Ca^{2+}
   d. I^{-}

48. Relative ionic sizes are indicated in Figure 12.9. Within a given horizontal row of the periodic chart, negative ions tend to be larger than positive ions because the negative ions contain a larger number of electrons in the valence shell. Within a vertical group of the periodic table, ionic size increases from top to bottom. In general, positive ions are smaller than the atoms they come from, whereas negative ions are larger than the atoms they come from.

   a. I^{-}
   b. Na^{+}
   c. Cs^{+}
   d. O

Chapter 12   Chemical Bonding   141

50. When atoms form covalent bonds, they try to attain a valence electronic configuration similar to that of the following noble gas element. When the elements in the first few horizontal rows of the periodic table form covalent bonds, they will attempt to gain configurations similar to the noble gases helium (2 valence electrons, duet rule), and neon and argon (8 valence electrons, octet rule).

52. These elements attain a total of eight valence electrons, making the valence electron configurations similar to those of the noble gases Ne and Ar.

54. When two atoms in a molecule are connected by a triple bond, we are indicating that the atoms share three pairs of electrons (6 electrons) in completing their outermost shells. A simple molecule containing a triple bond is acetylene, C2H2   H:C:::C:H

56. a.   Rb·

   b.   :C̈l·

   c.   :K̈r:

   d.   Ba:

   e.   ·P̈·

   f.   :Ät·

58. a.   C provides 4; each Br provides 7; total valence electrons = 32
   b.   N provides 5; each O provides 6; total valence electrons = 17
   c.   each C provides 4; each H provides 1; total valence electrons = 30
   d.   each O provides 6; each H provides 1; total valence electrons = 14

60. a. NH_3 N provides 5 valence electrons.
                Each H provides 1 valence electron.
                Total valence electrons = 8

$$H-\ddot{N}-H$$
$$|$$
$$H$$

142   Chapter 12   Chemical Bonding

    b. CI_4 C provides 4 valence electrons.
                      Each I provides 7 valence electrons.
                      Total valence electrons = 32

$$\begin{array}{c} \ddot{\text{:I:}} \\ | \\ \text{:}\ddot{\text{I}}\text{—C—}\ddot{\text{I}}\text{:} \\ | \\ \text{:}\ddot{\text{I}}\text{:} \end{array}$$

    c. $NCl3$ N provides 5 valence electrons.
                      Each Cl provides 7 valence electrons.
                      Total valence electrons = 26.

$$\begin{array}{c} \text{:}\ddot{\text{Cl}}\text{—}\ddot{\text{N}}\text{—}\ddot{\text{Cl}}\text{:} \\ | \\ \text{:}\ddot{\text{Cl}}\text{:} \end{array}$$

    d. $SiBr_4$ Si provides 4 valence electrons.
                      Each Br provides 7 valence electrons.
                      Total valence electrons = 32

$$\begin{array}{c} \text{:}\ddot{\text{Br}}\text{:} \\ | \\ \text{:}\ddot{\text{Br}}\text{—Si—}\ddot{\text{Br}}\text{:} \\ | \\ \text{:}\ddot{\text{Br}}\text{:} \end{array}$$

62. a. H_2S Each H provides 1 valence electron.
                      S provides 6 valence electrons.
                      Total valence electrons = 8

$$\text{H—}\ddot{\text{S}}\text{—H}$$

    b. SiF_4 Si provides 4 valence electrons.
                      Each F provides 7 valence electrons.
                  Total valence electrons = 32

$$\begin{array}{c} \text{:}\ddot{\text{F}}\text{:} \\ | \\ \text{:}\ddot{\text{F}}\text{—Si—}\ddot{\text{F}}\text{:} \\ | \\ \text{:}\ddot{\text{F}}\text{:} \end{array}$$

    c. C_2H_4 Each C provides 4 valence electrons.
                      Each H provides 1 valence electron.
                  Total valence electrons = 12

$$\begin{array}{c} \text{H}\quad\text{H} \\ |\quad\;| \\ \text{C}=\text{C} \\ |\quad\;| \\ \text{H}\quad\text{H} \end{array}$$

d. C₃H₈   Each C provides 4 valence electrons.
              Each H provides 1 valence electron.
              Total valence electrons = 20

$$\begin{array}{c} \text{H} \quad \text{H} \quad \text{H} \\ | \quad | \quad | \\ \text{H}-\text{C}-\text{C}-\text{C}-\text{H} \\ | \quad | \quad | \\ \text{H} \quad \text{H} \quad \text{H} \end{array}$$

64. a. NO₂   N provides 5 valence electrons.
              Each O provides 6 valence electrons.
              Total valence electrons = 17

$$\ddot{\text{O}}=\dot{\text{N}}-\ddot{\text{O}}: \qquad \ddot{\text{O}}=\dot{\text{N}}-\ddot{\text{O}}:$$

$$\cdot\ddot{\text{O}}-\dot{\text{N}}=\ddot{\text{O}} \qquad :\ddot{\text{O}}-\dot{\text{N}}=\ddot{\text{O}}$$

Note that there are some Lewis structures possible because of the odd number of electrons (the unpaired electron can be located on different atoms in the Lewis structure).

b. H₂SO₄   Each H provides 1 valence electron.
             S provides 6 valence electrons.
             Each O provides 6 valence electrons.
             Total valence electrons = 32

$$\begin{array}{c} :\ddot{\text{O}}: \\ | \\ \text{H}-\ddot{\text{O}}-\text{S}-\ddot{\text{O}}-\text{H} \\ | \\ :\ddot{\text{O}}: \end{array}$$

c. N₂O₄   Each N provides 5 valence electrons.
             Each O provides 6 valence electrons.
             Total valence electrons = 34

$$\begin{array}{cc} :\ddot{\text{O}}: \; :\ddot{\text{O}}: & :\text{O}: \; :\text{O}: \\ | \quad | & \| \quad \| \\ :\ddot{\text{O}}=\text{N}-\text{N}=\ddot{\text{O}}: & :\ddot{\text{O}}-\text{N}-\text{N}-\ddot{\text{O}}: \end{array}$$

144 Chapter 12 Chemical Bonding

66. a. ClO_3^-

Cl provides 7 valence electrons.
Each O provides 6 valence electrons.
The 1− charge means 1 additional electron.
Total valence electrons = 26

$$\left[\begin{array}{c} ::\!\ddot{Cl}\!-\!\ddot{O}\!-\!\ddot{Cl}\!:: \\ | \\ :\!\ddot{Cl}\!: \end{array} \right]^{1-}$$

b. O_2^{2-}

Each O provides 6 valence electrons.
The 2− charge means two additional valence electrons.
Total valence electrons = 14

$$\left[:\!\ddot{O}\!-\!\ddot{O}\!: \right]^{2-}$$

c. $C_2H_3O_2^-$

Each C provides 4 valence electrons.
Each H provides 1 valence electron.
The 1− charge means 1 additional valence electron.
Total valence electrons = 24

$$\left[\begin{array}{c} H \quad :O: \\ | \quad \| \\ H-C-C-\ddot{O}: \\ | \\ H \end{array} \right]^{1-} \quad \left[\begin{array}{c} H \quad :\ddot{O}: \\ | \quad | \\ H-C-C=\ddot{O} \\ | \\ H \end{array} \right]^{1-}$$

68. a. CN^-

C provides 4 valence electrons.
N provides 5 valence electrons.
The 1− charge means one additional valence electron.
Total valence electrons = 10

$$[:C\equiv N:]^-$$

b. HSO_4^-

H provides 1 valence electron.
S provides 6 valence electrons.
Each O provides 6 valence electrons.
The 1− charge means one additional valence electron.
Total valence electrons = 32

$$\left[\begin{array}{c} :\ddot{O}: \\ | \\ :\ddot{O}\!-\!S\!-\!\ddot{O}\!:H \\ | \\ :\ddot{O}: \end{array} \right]$$

c. N_3^- Each N provides 5 valence electrons.
The 1- charge means one additional valence electron.
Total valence electrons = 16

$$[:\ddot{N}=N=\ddot{N}:]^- \leftrightarrow [:N\equiv N-\ddot{\ddot{N}}:]^- \leftrightarrow [:\ddot{\ddot{N}}-N\equiv N:]^-$$

70. The geometric structure of NH_3 is that of a trigonal pyramid. The nitrogen atom of NH_3 is surrounded by four electron pairs (three are bonding, one is a lone pair). The H–N–H bond angle is somewhat less than 109.5° (due to the presence of the lone pair).

72. The geometric structure of CH_4 is that of a tetrahedron. The carbon atom of CH_4 is surrounded by four bonding electron pairs. The H–C–H bond angle is the characteristic angle of the tetrahedron, 109.5°.

74. The general molecular structure of a molecule is determined by *how many electron pairs* surround the central atom in the molecule, and by which of those electron pairs are used for *bonding* to the other atoms of the molecule. Nonbonding electron pairs on the central atom do, however, cause minor changes in the bond angles, compared to the ideal regular geometric structure.

76. You will remember from high school geometry, that two points in space are all that is needed to define a straight line. A diatomic molecule represents two points (the nuclei of the atoms) in space.

78. In NF_3, the nitrogen atom has *four* pairs of valence electrons, whereas in BF_3, there are only *three* pairs of valence electrons around the boron atom. The nonbonding electron pair on nitrogen in NF_3 pushes the three F atoms out of the plane of the N atom.

80. If you draw the Lewis structures for these species, you will see that each of the indicated atoms is surrounded by *four pairs* of electrons with a *tetrahedral* orientation. Although the electron pairs are arranged tetrahedrally, realize that the overall geometric shape of the molecules may *not* be tetrahedral, depending on what atoms are *bonded* to the four pairs of electrons on the central atom.

82. a. tetrahedral (four electron pairs on Ge, and four atoms attached)

b. trigonal pyramid (four electron pairs on P, but only three atoms attached to P)

c. nonlinear, V-shaped (four electron pairs on O, but only two atoms attached to O)

146   Chapter 12   Chemical Bonding

84. a. trigonal pyramid (four electron pairs on Cl, and four atoms attached to Cl)

   b. basically tetrahedral around P (four electron pairs on P, with four oxygen atoms attached to P; the hydrogen atom is attached to one of the oxygen atoms, and does not greatly distort the shape)

   c. trigonal pyramid (four electron pairs on S, but only three atoms attached to S)

86. a. approximately 109.5° (the molecule is V-shaped or nonlinear)

   b. approximately 109.5° (the molecule is trigonal pyramidal)

   c. 109.5°

   d. approximately 120° (the double bond makes the molecule flat)

88. In a covalent bond between two atoms of the same element, the electron pair is shared equally and the bond is nonpolar; with a bond between atoms of different elements, the electron pair is unequally shared and the bond is polar (assuming the elements have different electronegativities).

90. double

92. The bond with the larger electronegativity difference will be the more polar bond. See Figure 12.3 for electronegativities.

   a. S–F

   b. P–O

   c. C–H

94. break

96. In each case, the element *higher up* within a group of the periodic table has the higher electronegativity.

   a. Be

   b. N

   c. F

98. For a bond to be polar covalent, the atoms involved in the bond must have different electronegativities (must be of different elements).

   a. polar covalent (different elements)

   b. *non*polar covalent (two atoms of the same element)

c. *polar* covalent (different elements)

d. *nonpolar* covalent (atoms of the same element)

100. In a diatomic molecule containing two different elements, the more electronegative atom will be the negative end of the molecule, and the *less* electronegative atom will be the positive end.

    a. hydrogen

    b. iodine

    c. nitrogen

102. a. Al $1s^2\ 2s^2\ 2p^6\ 3s^2\ 3p^1$

       Al^{3+} $1s^2\ 2s^2\ 2p^6$

       Ne has the same configuration as Al^{3+}.

    b. Br $1s^2\ 2s^2\ 2p^6\ 3s^2\ 3p^6\ 4s^2\ 3d^{10}\ 4p^5$

       Br$^-$ $1s^2\ 2s^2\ 2p^6\ 3s^2\ 3p^6\ 4s^2\ 3d^{10}\ 4p^6$

       Kr has the same configuration as Br$^-$.

    c. Ca $1s^2\ 2s^2\ 2p^6\ 3s^2\ 3p^6\ 4s^2$

       Ca^{2+} $1s^2\ 2s^2\ 2p^6\ 3s^2\ 3p^6$

       Ar has the same configuration as Ca^{2+}.

    d. Li $1s^2\ 2s^1$

       Li$^+$ $1s^2$

       He has the same configuration as Li$^+$.

    e. F $1s^2\ 2s^2\ 2p^5$

       F$^-$ $1s^2\ 2s^2\ 2p^6$

       Ne has the same configuration as F$^-$.

104. a. Na$_2$Se   Na has one electron more than a noble gas; Se has two electrons fewer than a noble gas.

    b. RbF   Rb has one electron more than a noble gas; F has one electron less than a noble gas.

    c. K$_2$Te   K has one electron more than a noble gas; Te has two electrons fewer than a noble gas.

    d. BaSe   Ba has two electrons more than a noble gas; Se has two electrons fewer than a noble gas.

148   Chapter 12   Chemical Bonding

    e.   KAt   K has one electron more than a noble gas; At has one electron less than a noble gas.

    f.   FrCl   Fr has one electron more than a noble gas; Cl has one electron less than a noble gas.

106. Relative ionic sizes are indicated in Figure 12.9.

    a. Na^+

    b. Al^{3+}

    c. F^-

    d. Na^+

108. a.   Si provides 4; each H provides 1; total valence electrons = 8

       b.   S provides 6; each F provides 7; total valence electrons = 48

       c.   each N provides 5; each O provides 6; total valence electrons = 34

       d.   As provides 5; each I provides 7; total valence electrons = 26

110. a. N_2H_4 Each N provides 5 valence electrons.
Each H provides 1 valence electron.
Total valence electrons = 14

```
    ..  ..
H—N—N—H
    |   |
    H   H
```

       b. C_2H_6 Each C provides 4 valence electrons.
Each H provides 1 valence electron.
Total valence electrons = 14

```
    H   H
    |   |
H—C—C—H
    |   |
    H   H
```

       c. NCl_3 N provides 5 valence electrons.
Each Cl provides 7 valence electrons.
Total valence electrons = 26

```
        ..
:Cl—N—Cl:
      |
    :Cl:
      ..
```

d. SiCl₄   Si provides 4 valence electrons.
Each Cl provides 7 valence electrons.
Total valence electrons = 32

$$\begin{array}{c} :\!\ddot{\underset{..}{Cl}}\!: \\ | \\ :\!\ddot{\underset{..}{Cl}}\!-\!Si\!-\!\ddot{\underset{..}{Cl}}\!: \\ | \\ :\!\ddot{\underset{..}{Cl}}\!: \end{array}$$

112. a. NO₃⁻   N provides 5 valence electrons.
Each O provides 6 valence electrons.
The 1- charge means one additional valence electron.
Total valence electrons = 24

$$\left[\begin{array}{c} \ddot{O}\!=\!N\!-\!\ddot{\underset{..}{O}}\!: \\ | \\ :\!\ddot{\underset{..}{O}}\!: \end{array}\right]^{-} \leftrightarrow \left[\begin{array}{c} :\!\ddot{\underset{..}{O}}\!-\!N\!=\!\ddot{O} \\ | \\ :\!\ddot{\underset{..}{O}}\!: \end{array}\right]^{-} \leftrightarrow \left[\begin{array}{c} :\!\ddot{\underset{..}{O}}\!-\!N\!-\!\ddot{\underset{..}{O}}\!: \\ \| \\ :\ddot{O} \end{array}\right]^{-}$$

b. CO₃²⁻   C provides 4 valence electrons.
Each O provides 6 valence electrons.
The 2- charge means two additional valence electrons.
Total valence electrons = 24

$$\left[\begin{array}{c} \ddot{O}\!=\!C\!-\!\ddot{\underset{..}{O}}\!: \\ | \\ :\!\ddot{\underset{..}{O}}\!: \end{array}\right]^{2-} \leftrightarrow \left[\begin{array}{c} :\!\ddot{\underset{..}{O}}\!-\!C\!=\!\ddot{O} \\ | \\ :\!\ddot{\underset{..}{O}}\!: \end{array}\right]^{2-} \leftrightarrow \left[\begin{array}{c} :\!\ddot{\underset{..}{O}}\!-\!C\!-\!\ddot{\underset{..}{O}}\!: \\ \| \\ :\ddot{O} \end{array}\right]^{2-}$$

c. NH₄⁺   N provides 5 valence electrons.
Each H provides 1 valence electron.
The 1+ charge means one *less* valence electron.
Total valence electrons = 8

$$\left[\begin{array}{c} H \\ | \\ H\!-\!N\!-\!H \\ | \\ H \end{array}\right]^{+}$$

114. a. four electron pairs arranged tetrahedrally about C
    b. four electron pairs arranged tetrahedrally about Ge
    c. three electron pairs arranged trigonally (planar) around B

116. a. ClO_3^-, trigonal pyramid (lone pair on Cl)
 b. ClO_2^-, nonlinear (V-shaped, two lone pairs on Cl)
 c. ClO_4^-, tetrahedral (all pairs on Cl are bonding)

118. a. nonlinear (V-shaped)

 b. trigonal planar

 c. basically trigonal planar around the C (the H is attached to one of the O atoms, and distorts the shape around the carbon only slightly)

 d. linear

Cumulative Review: Chapters 11 and 12

2. An atom is said to be in its ground state when it is in its lowest possible energy state. When an atom possesses more energy than in its ground state, the atom is said to be in an excited state. An atom is promoted from its ground state to an excited state by absorbing energy; when the atom returns from an excited state to its ground state it emits the excess energy as electromagnetic radiation. Atoms do not gain or emit radiation randomly, but rather do so only in discrete bundles of radiation called photons. The photons of radiation emitted by atoms are characterized by the wavelength (color) of the radiation: longer wavelength photons carry less energy than shorter wavelength photons. The energy of a photon emitted by an atom corresponds exactly to the difference in energy between two allowed energy states in an atom: thus, we can use an observable phenomenon (emission of light by excited atoms), to gain insight into the energy changes taking place within the atom.

4. Bohr pictured the electron moving in only certain circular orbits around the nucleus. Each particular orbit (corresponding to a particular distance from the nucleus) had associated with it a particular energy (resulting from the attraction between the nucleus and the electron). When an atom absorbs energy, the electron moves from its ground state in the orbit closest to the nucleus ($n = 1$) to an orbit farther away from the nucleus ($n = 2, 3, 4, \ldots$). When an excited atom returns to its ground state, corresponding to the electron moving from an outer orbit to the orbit nearest the nucleus, the atom emits the excess energy as radiation. Since the Bohr orbits are of fixed distances from the nucleus and from each other, when an electron moves from one fixed orbit to another, the energy change is of a definite amount, which corresponds to a photon being emitted of a particular characteristic wavelength and energy. The original Bohr theory worked very well for hydrogen: Bohr even predicted emission wavelengths for hydrogen which had not yet been seen, which were subsequently found at the exact wavelengths Bohr had calculated. However, when the simple Bohr model for the atom was applied to the emission spectra of other elements, the theory could not predict or explain the observed emission spectra.

6. The lowest energy hydrogen atomic orbital is called the 1s orbital. The 1s orbital is spherical in shape (that is, the electron density around the nucleus is uniform in all directions from the nucleus). The 1s orbital represents a probability map of electron density around the nucleus for the first principal energy level. The orbital does not have a sharp edge (it appears fuzzy) since the probability of finding the electron does not drop off completely suddenly with distance from the nucleus. The orbital does not represent just a spherical surface on which the electron moves (this would be similar to Bohr's original theory): when we draw a picture to represent the 1s orbital we are indicating that the probability of finding the electron within this region of space is greater than 90%. We know that the likelihood of finding the electron within this orbital is very high, but we still don't know exactly where in this region the electron is at a given instant in time.

8. The third principal energy level of hydrogen is divided into three sublevels, the 3s, 3p, and 3d sublevels. The 3s subshell consists of the single 3s orbital: like the other s orbitals, the 3s orbital is spherical in shape. The 3p subshell consists of a set of three equal-energy 3p orbitals: each of these 3p orbitals has the same shape ("dumbbell"), but each of the 3p orbitals is oriented in a different direction in space. The 3d subshell consists of a set of five 3d orbitals: the 3d orbitals have the shapes indicated in Figure 11.23, and are oriented in different directions around the nucleus (students sometimes say that the 3d orbitals have the shape of a 4-leaf clover). The fourth principal energy level of hydrogen is dived into four sublevels, the 4s, 4p, 4d, and 4f orbitals. The 4s subshell consists of the single 4s orbital. The 4p subshell consists of a set of three 4p orbitals. The 4d subshell consists of a set of five 4d orbitals. The shapes of the 4s, 4p, and 4d orbitals are the same as the shapes of the orbitals of the third principal energy level (the orbitals of the fourth principal energy level are larger and further from the nucleus than the orbitals of the third level, however). The fourth principal energy level, because it is further from the nucleus, also contains a 4f subshell, consisting of seven 4f orbitals (the shapes of the 4f orbitals are beyond the scope of this text).

10. Atoms have a series of principal energy levels symbolized by the letter n. The $n = 1$ level is the closest to the nucleus, and the energies of the levels increase as the value of n increases going out from the nucleus. Each principal energy level is divided into a set of sublevels of different characteristic shapes (designated by the letters s, p, d, and f). Each sublevel is further subdivided into a set of orbitals: each s subshell consists of a single s orbital; each p subshell consists of a set of three p orbitals; each d subshell consists of a set of five d orbitals; etc. A given orbital can be empty or it can contain one or two electrons, but never more than two electrons (if an orbital contains two electrons, then the electrons must have opposite intrinsic spins). The shape we picture for an orbital represents only a probability map for finding electrons: the shape does not represent a trajectory or pathway for electron movements.

12. The valence electrons of an atoms are the electrons in the outermost shell of the atom. The core electrons are those in principal energy levels closer to the nucleus than the outermost shell (the core electrons are the electrons that are not valence electrons). Since the valence electrons are those in the outermost shell of the atom, it is these electrons that are affected by the presence of other atoms, and which are gained, lost, or shared with other atoms. The periodic table is basically arranged in terms of the valence electronic configurations of the elements (elements in the same vertical group have similar configurations): for example, Group 1 is the first column in the periodic table because all the elements in Group 1 have one valence electron. It is felt that the 4s subshell fills before the 3d subshell because the elements K and Ca have very similar properties to the other Group 1 and Group 2 elements (whereas the properties of Sc and Ti do not have such properties). For reasons beyond the scope of this book, it is

known conclusively that the 4s and 3d subshells are very close in energy, and it would not be unreasonable for the 4s subshell to fill first. Similarly, the 5s subshell fills before the 4d or 4f subshells.

14. The general periodic table you drew for Question 13 should be similar to that found in Figure 11.27 of the text. Just from the column and row location of an element, you should be able to determine what the valence shell of an element has for its electronic configuration. For example, the element in the third horizontal row, in the second vertical column, has $3s^2$ as its valence configuration. We know that the valence electrons are in the $n = 3$ shell because the element is in the third horizontal row. We know that the valence electrons are s electrons because the first two electrons in a horizontal row are always in an s subshell. We know that there are two electrons because the element is the second element in the horizontal row. As an additional example, the element in the seventh vertical column of the second horizontal row in the periodic table has valence configuration $2s^2 2p^5$.

16. The ionization energy of an atom represents the energy required to remove an electron from the atom. As one goes from top to bottom in a vertical group in the periodic table, the ionization energies decrease (it becomes easier to remove an electron). As one goes down within a group, the valence electrons are farther and farther from the nucleus and are less tightly held. The ionization energies increase when going from left to right within a horizontal row within the periodic table. The left-hand side of the periodic table is where the metallic elements are found, which lose electrons relatively easily. The right-hand side of the periodic table is where the nonmetallic elements are found: rather than losing electrons, these elements tend to gain electrons. Within a given horizontal row in the periodic table, the valence electrons are all in the same principal energy shell: however, as you go from left to right in the horizontal row, the nuclear charge which holds onto the electrons is increasing one unit with each successive element, making it that much more difficult to remove an electron. The relative sizes of atoms also vary systematically with the location of an element in the periodic table. Within a given vertical group, the atoms get progressively larger when going from the top of the group to the bottom: the valence electrons of the atoms are in progressively higher principal energy shells (and are progressively further from the nucleus) as we go down in a group. In going from left to right within a horizontal row in the periodic table, the atoms get progressively smaller. Although all the elements in a given horizontal row in the periodic table have their valence electrons in the same principal energy shell, the nuclear charge is progressively increasing from left to right, making the given valence shell progressively smaller as the electrons are drawn more closely to the nucleus.

18. Ionic bonding results when elements of very different electronegativities react with each other. Typically a metallic element reacts with a nonmetallic element, with the metallic element losing electrons and forming forming positive ions and the nonmetallic element gaining electrons and forming negative ions. The aggregate form of such

a compound consists of a crystal lattice of alternating positively and negatively charged ions. A given positive ion is attracted by several surrounding negatively charged ions, and a given negative ion is attracted by several surrounding positively charged ions. Similar electrostatic attractions go on in three dimensions throughout the crystal of ionic solid, leading to a very stable system (with very high melting and boiling points, for example). We know that ionically bonded solids do not conduct electricity in the solid state (since the ions are held tightly in place by all the attractive forces), but such substances are strong electrolytes when melted or when dissolved in water (either of which process sets the ions free to move around). There are many examples of ionic substances: for example, when sodium metal is placed in a container of chlorine gas, an explosion takes place (a very visual indicator of the strength of the ionic bond) and a white deposit of sodium chloride forms.

$$2Na(s) + Cl_2(g) \rightarrow 2Na^+Cl^-(s)$$

20. Electronegativity represents the relative ability of an atom in a molecule to attract shared electrons towards itself. In order for a bond to be polar, one of the atoms in the bond must attract the shared electron pair towards itself and away from the other atom of the bond: this can only happen if one atom of the bond is more electronegative than the other (that is, that there is a considerable difference in electronegativity for the two atoms of the bond). If two atoms in a bond have the same electronegativity, then the two atoms pull the electron pair equally and the bond is nonpolar covalent. If two atoms sharing a pair of electrons have vastly different electronegativities, the electron pair will be pulled so strongly by the more electronegative atom that a negative ion may be formed (as well as a positive ion for the second atom) and ionic bonding will result. If the difference in electronegativity between two atoms sharing an electron pair is somewhere in between these two extremes (equal sharing of the electron pair and formation of ions), then a polar covalent bond results.

22. It has been observed over many, many experiments that when an active metal like sodium or magnesium reacts with a nonmetal, the sodium atoms always form Na^+ ions and the magnesium atoms always form Mg^{2+} ions. It has been further observed that aluminum always forms only the Al^{3+} ion, and that when nitrogen, oxygen, or fluorine form simple ions, the ions that are formed are always N^{3-}, O^{2-}, and F^-, respectively. Clearly the facts that these elements always form the same ions and that those ions all contain eight electrons in the outermost shell, led scientists to speculate that there must be something very fundamentally stable about a species that has eight electrons in its outermost shell (like the noble gas neon). The repeated observation that so many elements, when reacting, tend to attain an electronic configuration that is isoelectronic with a noble gas led chemists to speculate that all elements try to attain such a configuration for their outermost shells. In general, when atoms of a metal react with atoms of a nonmetal, the metal atoms lose electrons until they have the configuration of the preceding noble gas, and the nonmetal atoms gain electrons until they

have the configuration of the following noble gas. Covalently and polar covalently bonded molecules also strive to attain pseudo-noble gas electronic configurations. For a covalently bonded molecule like F_2, in which neither fluorine atom has a greater tendency than the other to gain or lose electrons completely, each F atom provides one electron of the pair of electrons which constitutes the covalent bond. Each F atom feels also the influence of the other F atom's electron in the shared pair, and each F atom effectively fills its outermost shell. Similarly, in polar covalently bonded molecules like HF or HCl, the shared pair of electrons between the atoms effectively completes the outer electron of each atom simultaneously to give each atom a noble gas-like electronic configuration.

24. Bonding between atoms to form a molecule involves only the valence electrons of the atoms(not the inner core electrons. So when we draw the Lewis structure of a molecule, we show only these valence electrons (both bonding valence electrons and non-bonding valence electrons, however). The most important requisite for the formation of a stable compound, and which we try to demonstrate when we write Lewis structures, is that each atom of a molecule attains a noble gas electron configuration. When we write Lewis structures, we arrange the bonding and nonbonding valence electrons to try to complete the octet (or duet) for as many atoms as is possible.

26. Obviously, you could choose practically any molecules for your discussion. Let's illustrate the method for ammonia, NH_3. First count up the total number of valence electrons available in the molecule (without regard to what atom they officially come from); remember that for the representative elements, the number of valence electrons is indicated by what group the element is found in on the periodic table. For NH_3, since nitrogen is in Group 5, one nitrogen atom would contribute five valence electron. Since hydrogen atoms only have one electron each, the three hydrogen atoms provide an additional three valence electrons, for a total of eight valence electrons overall. Next write down the symbols for the atoms in the molecule, and use one pair of electrons (represented by a line) to form a bond between each pair of bound atoms.

$$H-N-H$$
$$|$$
$$H$$

These three bonds use six of the eight valence electrons. Since each hydrogen already has its duet in what we have drawn so far, while the nitrogen atom only has six electrons around it so far, the final two valence electrons must represent a lone pair on the nitrogen.

$$H-\ddot{N}-H$$
$$|$$
$$H$$

28. There are several types of exceptions to the octet rule described in the text. The octet rule is really a "rule of thumb" which we apply to molecules unless we have some evidence that a molecule does not follow the rule. There are some common molecules which, from experimental measurements, we know do not follow the octet rule. Boron and beryllium compounds sometimes do not fit the octet rule. For example, in BF_3, the boron atom only has six valence electrons in its outermost shell, whereas in BeF_2, the beryllium atom only has four electrons in its outermost shell. Other molecules which are exceptions to the octet rule include any molecule with an odd number of valence electrons (such as NO or NO_2): you can't get an octet (an even number) of electrons around each atom in a molecule with an odd number of valence electrons. Even the oxygen gas we breathe is an exception to the octet rule: although we can write a Lewis structure for O_2 satisfying the octet rule for each oxygen, we know from experiment that O_2 contains unpaired electrons.

30.

Number of valence pairs	bond angle	example(s)
2	180°	BeF_2, BeH_2
3	120°	BCl_3
4	109.5°	CH_4, CCl_4, GeF_4

32. The general shape of the orbitals in a particular type of subshell (s, p, or d) is the same regardless of the principal energy level (n value). Thus the 2p, 3p, and 4p orbitals all have basically the same shape (dumbbells), however the 4p orbitals are larger than the 3p orbitals, which in turn are larger than the 2p orbitals. The individual p orbitals within a given subshell are oriented in different directions. 90° apart from one another. Similarly, the 1s, 2s and 3s orbitals are all spherical, with the 3s orbital the largest and the 1s orbital the smallest (students sometimes like to picture the orbitals like the layers in an onion, with the 1s orbital inside the 2s orbital, which is inside the 3s orbital). The 3d orbitals (along with the general shapes of the s and p orbitals) are shown in Figures 11.21, 11.22, and 11.23.

34.

Group	General configuration
1	[noble gas]ns^1
2	[noble gas]ns^2
3	[noble gas]ns^2np^1
4	[noble gas]ns^2np^2
5	[noble gas]ns^2np^3
6	[noble gas]ns^2np^4
7	[noble gas]ns^2np^5
8	[noble gas]ns^2np^6 (itself a noble gas)

36. a. GeCl$_4$

$$\begin{array}{c} \ddot{\text{:}\ddot{\text{Cl}}\text{:}} \\ | \\ \text{:}\ddot{\text{Cl}}-\text{Ge}-\ddot{\text{Cl}}\text{:} \\ | \\ \text{:}\ddot{\text{Cl}}\text{:} \end{array}$$

b. NO$_3^-$

$$\left[\ddot{\text{O}}=\text{N}-\ddot{\ddot{\text{O}}}\text{:} \atop \underset{\ddot{\text{:O:}}}{|}\right]^- \leftrightarrow \left[\text{:}\ddot{\ddot{\text{O}}}-\text{N}=\ddot{\text{O}} \atop \underset{\ddot{\text{:O:}}}{|}\right]^- \leftrightarrow \left[\text{:}\ddot{\ddot{\text{O}}}-\text{N}-\ddot{\ddot{\text{O}}}\text{:} \atop \underset{\ddot{\text{:O:}}}{||}\right]^-$$

c. SO$_4^{2-}$

$$\left[\begin{array}{c} \ddot{\text{:O:}} \\ | \\ \text{:}\ddot{\text{O}}-\text{S}-\ddot{\text{O}}\text{:} \\ | \\ \ddot{\text{:O:}} \end{array}\right]^{2-}$$

d. SO$_3^{2-}$

$$\left[\text{:}\ddot{\ddot{\text{O}}}-\text{S}-\ddot{\ddot{\text{O}}}\text{:} \atop \underset{\ddot{\text{:O:}}}{|}\right]^{2-}$$

e. SO$_2$

$$\ddot{\text{O}}=\ddot{\text{S}}-\ddot{\ddot{\text{O}}}\text{:} \leftrightarrow \text{:}\ddot{\ddot{\text{O}}}-\ddot{\text{S}}=\ddot{\text{O}}$$

f. O$_2$

$$\ddot{\mathrm{O}}=\ddot{\mathrm{O}}$$

g. O$_3$

$$\ddot{\mathrm{O}}=\ddot{\mathrm{O}}-\ddot{\mathrm{O}}: \leftrightarrow :\ddot{\mathrm{O}}-\ddot{\mathrm{O}}=\ddot{\mathrm{O}}$$

h. CH$_3$NH$_2$

$$\begin{array}{c} \mathrm{H} \\ | \\ \mathrm{H}-\mathrm{C}-\ddot{\mathrm{N}}-\mathrm{H} \\ | \quad | \\ \mathrm{H} \quad \mathrm{H} \end{array}$$

i. C$_2$H$_6$

$$\begin{array}{c} \mathrm{H} \quad \mathrm{H} \\ | \quad | \\ \mathrm{H}-\mathrm{C}-\mathrm{C}-\mathrm{H} \\ | \quad | \\ \mathrm{H} \quad \mathrm{H} \end{array}$$

j. C$_2$H$_4$

$$\begin{array}{c} \mathrm{H} \quad \mathrm{H} \\ | \quad | \\ \mathrm{C}=\mathrm{C} \\ | \quad | \\ \mathrm{H} \quad \mathrm{H} \end{array}$$

k. BF$_3$

$$:\ddot{\underset{..}{F}}-B-\ddot{\underset{..}{F}}:$$
$$|$$
$$:\ddot{\underset{..}{F}}:$$

l. BeF$_2$

$$:\ddot{\underset{..}{F}}-Be-\ddot{\underset{..}{F}}:$$

m. NO$_2^-$

$$\left[:\ddot{\underset{..}{O}}-\ddot{N}=\ddot{O}\right]^- \leftrightarrow \left[\ddot{O}=\ddot{N}-\ddot{\underset{..}{O}}:\right]^-$$

n. PF$_3$

$$:\ddot{\underset{..}{F}}-\ddot{P}-\ddot{\underset{..}{F}}:$$
$$|$$
$$:\ddot{\underset{..}{F}}:$$

o. N$_2$H$_4$

$$H-\ddot{N}-\ddot{N}-H$$
$$|\ \ |$$
$$H\ \ H$$

Chapter 13   Gases

2. Solids and liquids have essentially fixed volumes and are not able easily to be compressed. Gases have volumes that depend on their conditions, and can be compressed or expanded by changes in those conditions. Although the particles of matter in solids are essentially fixed in position (the solid is rigid), the particles in liquids and gases are free to move.

4. Atmospheric pressure is most commonly measured with a mercury barometer (the pressure of the atmosphere is sufficient to maintain the height of a column of mercury to approximately 76 cm at sea level). The pressure of the atmosphere varies on the surface of the earth due to weather conditions, and also varies with altitude.

6. 760.

8. a. $110.2 \text{ kPa} \times \dfrac{10^3 \text{ Pa}}{1 \text{ kPa}} \times \dfrac{1 \text{ atm}}{101,325 \text{ Pa}} = 1.088 \text{ atm}$

    b. $74.2 \text{ cm Hg} \times \dfrac{10 \text{ mm Hg}}{1 \text{ cm Hg}} \times \dfrac{1 \text{ atm}}{760 \text{ mm Hg}} = 0.976 \text{ atm}$

    c. $441 \text{ mm Hg} \times \dfrac{1 \text{ atm}}{760 \text{ mm Hg}} = 0.580 \text{ atm}$

    d. $921 \text{ torr} \times \dfrac{1 \text{ atm}}{760 \text{ torr}} = 1.21 \text{ atm}$

10. a. $1.02 \text{ atm} \times \dfrac{760 \text{ mm Hg}}{1 \text{ atm}} = 775 \text{ mm Hg}$

    b. $121.4 \text{ kPa} \times \dfrac{10^3 \text{ Pa}}{1 \text{ kPa}} \times \dfrac{760 \text{ mm Hg}}{101,325 \text{ Pa}} = 911 \text{ mm Hg}$

    c. 792 torr = 792 mm Hg (1 torr = 1 mm Hg)

    d. $1.09 \times 10^4 \text{ Pa} \times \dfrac{760 \text{ mm Hg}}{101,325 \text{ Pa}} = 81.8 \text{ mm Hg}$

12. a. $2.07 \times 10^6 \text{ Pa} \times \dfrac{1 \text{ kPa}}{10^3 \text{ Pa}} = 2.07 \times 10^3 \text{ kPa}$

    b. $795 \text{ mm Hg} \times \dfrac{101,325 \text{ Pa}}{760 \text{ mm Hg}} \times \dfrac{1 \text{ kPa}}{10^3 \text{ Pa}} = 106 \text{ kPa}$

    c. $10.9 \text{ atm} \times \dfrac{101,325 \text{ Pa}}{1 \text{ atm}} \times \dfrac{1 \text{ kPa}}{10^3 \text{ Pa}} = 1.10 \times 10^3 \text{ kPa}$

    d. $659 \text{ torr} \times \dfrac{101,325 \text{ Pa}}{760 \text{ torr}} \times \dfrac{1 \text{ kPa}}{10^3 \text{ Pa}} = 87.9 \text{ kPa}$

Chapter 13   Gases   161

14. increases

16. $PV = k$; $P_1V_1 = P_2V_2$

18. a. $P_1 = 1.00$ atm $P_2 = 699$ torr $= 0.920$ atm
        $V_1 = 541$ mL $V_2 = ?$ mL

    $V_2 = \dfrac{P_1V_1}{P_2} = \dfrac{(1.00 \text{ atm})(541 \text{ mL})}{(0.920 \text{ atm})} = 588$ mL

    b. $P_1 = 110.2$ kPa $P_2 = 0.995$ atm $= 100.8$ kPa
        $V_1 = 2.32$ L $V_2 = ?$ L

    $V_2 = \dfrac{P_1V_1}{P_2} = \dfrac{(110.2 \text{ kPa})(2.32 \text{ L})}{(100.8 \text{ kPa})} = 2.53$ L

    c. $P_1 = 135$ atm $= 1.026 \times 10^5$ mm Hg $P_2 = ?$ mm Hg
        $V_1 = 4.15$ mL $V_2 = 10.0$ mL

    $P_2 = \dfrac{P_1V_1}{V_2} = \dfrac{(1.026 \times 10^5 \text{ mm Hg})(4.15 \text{ mL})}{(10.0 \text{ mL})} = 4.26 \times 10^4$ mm Hg

20. a. $P_1 = 1.82$ atm $P_2 = 245$ mm Hg $= 0.322$ atm
        $V_1 = 561$ mL $V_2 = ?$ mL

    $V_2 = \dfrac{P_1V_1}{P_2} = \dfrac{(1.82 \text{ atm})(561 \text{ mL})}{(0.322 \text{ atm})} = 3.17 \times 10^3$ mL

    b. $P_1 = 1.82$ atm $P_2 = 1.82$ kPa $= 0.0180$ atm
        $V_1 = 561$ mL $V_2 = ?$ mL

    $V_2 = \dfrac{P_1V_1}{P_2} = \dfrac{(1.82 \text{ atm})(561 \text{ mL})}{(0.0180 \text{ atm})} = 5.67 \times 10^4$ mL

    c. $P_1 = 1.82$ atm $P_2 = 2.45 \times 10^4$ Pa $= 0.242$ atm
        $V_1 = 561$ mL $V_2 = ?$ mL

    $V_2 = \dfrac{P_1V_1}{P_2} = \dfrac{(1.82 \text{ atm})(561 \text{ mL})}{(0.242 \text{ atm})} = 4.22 \times 10^3$ mL

22. If the pressure exerted on the gas in the balloon is decreased, the volume of the gas in the balloon will increase in inverse proportion to the factor by which the pressure was changed. The factor in this example is (1.01 atm/0.562 atm) = 1.80

24. $P_1 = 760$ mm $= 1.00$ atm $P_2 = ?$ atm
    $V_1 = 1.00$ L $V_2 = 50.0$ mL $= 0.0500$ L

    $P_2 = \dfrac{P_1V_1}{V_2} = \dfrac{(1.00 \text{ atm})(1.00 \text{ L})}{(0.0500 \text{ L})} = 20.0$ atm

162   Chapter 13   Gases

26. Charles's law states that the volume of an ideal gas sample varies linearly with the absolute temperature of the gas sample. An experiment such as those indicated in Figures 13.6 and 13.7 can be performed to determine absolute zero. The volume of a sample of gas is measured at several convenient temperatures (e.g., between 0° and 100° C) and the data is plotted. The straight line obtained is then extrapolated to the point where the volume of the gas would become zero. The temperature at which the volume of the gas would be predicted to become zero is then absolute zero.

28. $V = bT$; $V_1/T_1 = V_2/T_2$

30. $V_1 = 525$ mL $\qquad\qquad\qquad\qquad V_2 = ?$ mL
    $T_1 = 25$ °C $= 298$ K $\qquad\qquad T_2 = 50$ °C $= 323$ K

    $V_2 = \dfrac{V_1 T_2}{T_1} = \dfrac{(525 \text{ mL})(323 \text{ K})}{(298 \text{ K})} = 569$ mL

32. a. $V_1 = 25.0$ L $\qquad\qquad\qquad V_2 = 50.0$ L
        $T_1 = 0$ °C $= 273$ K $\qquad\qquad T_2 = ?$ K

    $T_2 = \dfrac{V_2 T_1}{V_1} = \dfrac{(50.0 \text{ L})(273 \text{ K})}{(25.0 \text{ L})} = 546$ K $= 273$ °C

    b. $V_1 = 247$ mL $\qquad\qquad\qquad V_2 = 255$ mL
        $T_1 = 25$ °C $= 298$ K $\qquad\qquad T_2 = ?$ K

    $T_2 = \dfrac{V_2 T_1}{V_1} = \dfrac{(255 \text{ mL})(298 \text{ K})}{(247 \text{ mL})} = 308$ K $= 35$ °C

    c. $V_1 = 1.00$ mL $\qquad\qquad\qquad V_2 = ?$ mL
        $T_1 = -272$ °C $= 1$ K $\qquad\qquad T_2 = 25$ °C $= 298$ K

    $V_2 = \dfrac{V_1 T_2}{T_1} = \dfrac{(1.00 \text{ mL})(298 \text{ K})}{(1 \text{ K})} = 298$ mL $= 3 \times 10^2$ mL

34. a. $V_1 = 2.01 \times 10^2$ L $\qquad\qquad V_2 = 5.00$ L
        $T_1 = 1150$ °C $= 1423$ K $\qquad T_2 = ?$ K

    $T_2 = \dfrac{V_2 T_1}{V_1} = \dfrac{(5.00 \text{ L})(1423 \text{ K})}{(2.01 \times 10^2 \text{ L})} = 35.4$ K $= -238$ °C

    b. $V_1 = 44.2$ mL $\qquad\qquad\qquad V_2 = ?$ mL
        $T_1 = 298$ K $\qquad\qquad\qquad\qquad T_2 = 0$ K

    $V_2 = \dfrac{V_1 T_2}{T_1} = \dfrac{(44.2 \text{ L})(0 \text{ K})}{(298 \text{ K})} = 0$ mL (0 K is absolute zero)

    c. $V_1 = 44.2$ mL $\qquad\qquad\qquad V_2 = ?$ mL
        $T_1 = 298$ K $\qquad\qquad\qquad\qquad T_2 = 0$ °C $= 273$ K

$$V_2 = \frac{V_1 T_2}{T_1} = \frac{(44.2 \text{ mL})(273 \text{ K})}{(298 \text{ K})} = 40.5 \text{ mL}$$

36. You should be able to answer these without having to set up a formal calculation. Charles's law says that the volume of a gas sample is *directly* proportional to its absolute temperature. So if a sample of neon has a volume of 266 mL at 25.2 °C (298 K), then the volume will become half as big at half the absolute temperature (149 K). The volume of the gas sample will become twice as big at twice the absolute temperature (596 K, 323°C).

38. One method of solution (using Charles's Law in the form $V_1/T_1 = V_2/T_2$) is shown in Example 13.6. A second method might make use of Charles's Law in the form $V = kT$. Using the information that the gas thermometer has a volume of 135 mL at 11°C (284 K), we can solve for the value of the proportionality constant k in the formula, and then use this information to calculate the additional temperatures requested.

    $V = kT$

    $$k = \frac{V}{T} = \frac{135 \text{ mL}}{284 \text{ K}} = 0.475 \text{ mL/K}$$

    $$T = \frac{V}{k} = \frac{V}{0.475 \text{ mL/K}} = 2.11V$$

    For 113 mL, $T = 2.11(113) = 238$ K (−35°C)

    For 142 mL, $T = 2.11(142) = 299$ K (26°C)

    For 155 mL, $T = 2.11(155) = 326$ K (53°C)

    For 127 mL, $T = 2.11(127) = 267$ K (−6°C)

40. $V = an$; $V_1/n_1 = V_2/n_2$

42. $V_1 = 11.2$ L $V_2 = ?$ L
    $n_1 = 0.500$ mol $n_2 = 2.00$ mol

    $$V_2 = \frac{V_1 n_2}{n_1} = \frac{(11.2 \text{ L})(2.00 \text{ mol})}{(0.500 \text{ L})} = 44.8 \text{ L}$$

44. $V_1 = 100.$ L $V_2 = ?$ L
    $n_1 = 46.2$ g/32.00 g mol^{-1} $n_2 = 5.00$ g/32.00 g mol^{-1}

    $$V_2 = \frac{V_2 n_2}{n_1} = \frac{(100. \text{ L})(5.00 \text{ g}/32.00 \text{ g mol}^{-1})}{(46.2 \text{ g}/32.00 \text{ g mol}^{-1})} = 10.8 \text{ L}$$

    Note that the *molar mass* of the O_2 gas cancels out in this calculation. Since the number of moles of Ne (or any gas) present in a sample is *directly proportional* to the mass of the gas sample, the problem could also have been set up directly in terms of the masses.

164   Chapter 13   Gases

46. Real gases most closely approach ideal gas behavior under conditions of relatively high temperatures (0°C or higher) and relatively low pressures (1 atm or lower).

48. For an ideal gas, $PV = nRT$ is true under any conditions. Consider a particular sample of gas (so that n remains constant) at a particular fixed pressure (so that P remains constant also). Suppose that at temperature T_1 the volume of the gas sample is V_1. Then for this set of conditions, the ideal gas equation would be given by

$$PV_1 = nRT_1$$

If we then change the temperature of the gas sample to a new temperature T_2, the volume of the gas sample changes to a new volume V_2. For this new set of conditions, the ideal gas equation would be given by

$$PV_2 = nRT_2$$

If we make a ratio of these two expressions for the ideal gas equation for this gas sample, and cancel out terms that are constant for this situation (P, n, and R) we get

$$\frac{PV_1}{PV_2} = \frac{nRT_1}{nRT_2}$$

$$\frac{V_1}{V_2} = \frac{T_1}{T_2}$$

which can be rearranged to the familiar form of Charles's law

$$\frac{V_1}{T_1} = \frac{V_2}{T_2}$$

50. a. $P = 782$ mm Hg $= 1.03$ atm

   $T = 27$ °C $= 300$ K

   $$V = \frac{nRT}{P} = \frac{(0.210 \text{ mol})(0.08206 \text{ L atm mol}^{-1} \text{ K}^{-1})(300 \text{ K})}{(1.03 \text{ atm})} = 5.02 \text{ L}$$

   b. $V = 644$ mL $= 0.644$ L

   $$P = \frac{nRT}{V} = \frac{(0.0921 \text{ mol})(0.08206 \text{ L atm mol}^{-1} \text{ K}^{-1})(303 \text{ K})}{(0.644 \text{ L})} = 3.56 \text{ atm}$$

   $P = 3.56$ atm $= 2.70 \times 10^3$ mm Hg

   c. $P = 745$ mm $= 0.980$ atm

   $$T = \frac{PV}{nR} = \frac{(0.980 \text{ atm})(11.2 \text{ L})}{(0.401 \text{ mol})(0.08206 \text{ L atm mol}^{-1} \text{ K}^{-1})} = 334 \text{ K}$$

52. a. $T = 25$ °C $= 298$ K

   $$V = \frac{(0.00831 \text{ mol})(0.08206 \text{ L atm mol}^{-1} \text{ K}^{-1})(298 \text{ K})}{(1.01 \text{ atm})} = 0.201 \text{ L}$$

b. $V = 602$ mL $= 0.602$ L

$$P = \frac{(8.01 \times 10^{-3} \text{ mol})(0.08206 \text{ L atm mol}^{-1} \text{ K}^{-1})(310 \text{ K})}{(0.602 \text{ L})} = 0.338 \text{ atm}$$

c. $V = 629$ mL $= 0.629$ L

$T = 35$ °C $= 308$ K

$$n = \frac{(0.998 \text{ atm})(0.629 \text{ L})}{(0.08206 \text{ L atm mol}^{-1} \text{ K}^{-1})(308 \text{ K})} = 2.48 \times 10^{-2} \text{ mol}$$

54. Molar mass of $O_2 = 32.00$ g

$$5.03 \text{ g } O_2 \times \frac{1 \text{ mol}}{32.00 \text{ g}} = 0.157 \text{ mol } O_2$$

$T = 28$ °C $= 301$ K

$$V = \frac{nRT}{P} = \frac{(0.157 \text{ mol})(0.08206 \text{ L atm mol}^{-1} \text{ K}^{-1})(301 \text{ K})}{(0.998 \text{ atm})} = 3.89 \text{ L}$$

56. The number of moles of *any* ideal gas that can be contained in the tank under the given conditions can first be calculated.

$T = 24$ °C $= 297$ K

$$n = \frac{PV}{RT} = \frac{(135 \text{ atm})(200 \text{ L})}{(0.08206 \text{ L atm mol}^{-1} \text{ K}^{-1})(297 \text{ K})} = 1.11 \times 10^3 \text{ mol gas}$$

Molar masses: He, 4.003 g; H_2, 2.016 g

for He: $1.11 \times 10^3 \text{ mol He} \times \dfrac{4.003 \text{ g}}{1 \text{ mol}} = 4.44 \times 10^3 \text{ g He} = 4.44 \text{ kg He}$

for H_2: $1.11 \times 10^3 \text{ mol } H_2 \times \dfrac{2.016 \text{ g}}{1 \text{ mol}} = 2.24 \times 10^3 \text{ g } H_2 = 2.24 \text{ kg } H_2$

58. Molar mass of $N_2 = 28.02$ g

$$16.3 \text{ g } N_2 \times \frac{1 \text{ mol}}{28.02 \text{ g}} = 0.582 \text{ mol } N_2$$

$$T = \frac{PV}{nR} = \frac{(1.25 \text{ atm})(25.0 \text{ L})}{(0.582 \text{ mol})(0.08206 \text{ L atm mol}^{-1} \text{ K}^{-1})} = 654 \text{ K} = 381 \text{ °C}$$

60. Molar mass of Ar $= 39.95$ g

$$51.3 \text{ lb} \times \frac{453.54 \text{ g}}{1 \text{ lb}} \times \frac{1 \text{ mol}}{39.95 \text{ g}} = 582 \text{ mol Ar}$$

$T = 25$ °C $= 298$ K

$$P = \frac{nRT}{V} = \frac{(582 \text{ mol})(0.08206 \text{ L atm mol}^{-1} \text{ K}^{-1})(298 \text{ K})}{(212 \text{ L})} = 67.2 \text{ atm}$$

166   Chapter 13   Gases

62. $P_1 = 0.981$ atm $\qquad\qquad P_2 = 1.15$ atm
     $V_1 = 125$ mL $\qquad\qquad V_2 = ?$ mL
     $T_1 = 100\ °C = 373$ K $\qquad T_2 = 25\ °C = 298$ K

   $V_2 = \dfrac{P_1 V_1 T_2}{P_2 T_1} = \dfrac{(0.981\text{ atm})(125\text{ mL})(298\text{ K})}{(1.15\text{ atm})(373\text{ K})} = 85.2$ mL

64. Molar mass of H_2 = 2.016 g

   $5.00\text{ g }H_2 \times \dfrac{1\text{ mol}}{2.016\text{ g}} = 2.48$ mol H_2

   $P = 761$ mm $= 1.001$ atm

   $T = \dfrac{PV}{nR} = \dfrac{(1.001\text{ atm})(50.0\text{ L})}{(2.48\text{ mol})(0.08206\text{ L atm mol}^{-1}\text{ K}^{-1})} = 246\text{ K} = -27\ °C$

66. As a gas is bubbled through water, the bubbles of gas become saturated with water vapor, thus forming a gaseous mixture. The total pressure in a sample of gas which has been collected by bubbling through water is made up of two components: the pressure of the gas of interest and the pressure of water vapor. The partial pressure of the gas of interest is then the total pressure of the sample minus the vapor pressure of water.

68. The volume of a sample of an ideal gas (at a given temperature and pressure) depends on the total number of moles of gas present, not on the specific identity of the gas. For a gaseous mixture as in this problem, the volume of the gas will depend on the total *combined* moles of nitrogen and oxygen.

   Molar masses: N_2, 28.02 g; O_2, 32.00 g

   $T = 25\ °C = 298$ K

   $5.23\text{ g }N_2 \times \dfrac{1\text{ mol}}{28.02\text{ g}} = 0.1867$ mol N_2

   $4.41\text{ g }O_2 \times \dfrac{1\text{ mol}}{32.00\text{ g}} = 0.1378$ mol O_2

   $n_{total} = 0.1867\text{ mol} + 0.1378\text{ mol} = 0.3245\text{ mol} = 0.325$ mol

   $V = \dfrac{nRT}{P} = \dfrac{(0.325\text{ mol})(0.08206\text{ L atm mol}^{-1}\text{ K}^{-1})(298\text{ K})}{(1.00\text{ atm})} = 7.94$ L

70. Molar masses: O_2, 32.00 g; N_2, 28.02 g

   $50.1\text{ g }O_2 \times \dfrac{1\text{ mol}}{32.00\text{ g}} = 1.566$ mol O_2

   $26.3\text{ g }N_2 \times \dfrac{1\text{ mol}}{28.02\text{ g}} = 0.9386$ mol N_2

   $n_{total} = 1.566\text{ mol} + 0.9386\text{ mol} = 2.505$ mol

Chapter 13   Gases   167

$T = 27 \,°C = 300 \text{ K}$

$P = \dfrac{nRT}{V} = \dfrac{(2.505 \text{ mol})(0.08206 \text{ L atm mol}^{-1} \text{ K}^{-1})(300 \text{ K})}{(25.0 \text{ L})} = 2.47 \text{ atm}$

72. The pressures must be expressed in the same units, either mm Hg or atm.

    $P_{oxygen} = P_{total} - P_{water\ vapor}$

    1.02 atm = 775 mm Hg

    P_{oxygen} = 775 mm Hg − 23.756 mm Hg = 751.244 mm = 751 mm Hg

74. 1.032 atm = 784.3 mm Hg

    $P_{hydrogen}$ = 784.3 mm Hg − 32 mm Hg = 752.3 mm Hg = 0.990 atm

    240 mL = 0.240 L

    30°C + 273 = 303 K

    $n_{hydrogen} = P_{hydrogen} V/RT = \dfrac{(0.990 \text{ atm})(0.240 \text{ L})}{(0.08206 \text{ L atm mol}^{-1} \text{ K}^{-1})(303 \text{ K})} = 0.00956 \text{ mol}$

    $0.00956 \text{ mol H}_2 \times \dfrac{1 \text{ mol Zn}}{1 \text{ mol H}_2} = 0.00956$ mol of Zn must have reacted

    molar mass of Zn = 65.38 g

    $0.00956 \text{ mol Zn} \times \dfrac{65.38 \text{ g Zn}}{1 \text{ mol Zn}} = 0.625$ g Zn must have reacted

76. A theory is successful if it explains known experimental observations. Theories which have been successful in the past may not be successful in the future (for example, as technology evolves, more sophisticated experiments may be possible in the future).

78. collisions

80. no

82. If the temperature of a sample of gas is increased, the average kinetic energy of the particles of gas increases. This means that the speeds of the particles increase. If the particles have a higher speed, they will hit the walls of the container more frequently and with greater force, thereby increasing the pressure.

84. Standard Temperature and Pressure, STP = 0 °C, 1 atm pressure. These conditions were chosen because they are easy to attain and reproduce *experimentally*. The barometric pressure within a laboratory is likely to be near 1 atm most days, and 0°C can be attained with a simple ice bath.

86. Molar mass of P_4 = 123.88 g

    $2.51 \text{ g P}_4 \times \dfrac{1 \text{ mol}}{123.88 \text{ g}} = 0.02026 \text{ mol P}_4$

168   Chapter 13   Gases

From the balanced chemical equation, the amount of hydrogen needed is

$$0.02026 \text{ mol } P_4 \times \frac{6 \text{ mol } H_2}{1 \text{ mol } P_4} = 0.1216 \text{ mol } H_2$$

$T = 25 \,°C = 298 \text{ K}$ $\qquad$ $P = 753 \text{ mm Hg} = 0.991 \text{ atm}$

$$V = \frac{nRT}{P} = \frac{(0.122 \text{ mol})(0.08206 \text{ L atm mol}^{-1} \text{ K}^{-1})(298 \text{ K})}{(0.991 \text{ atm})} = 3.00 \text{ L } H_2$$

88. Molar mass of Zn = 65.38 g

$$1.13 \text{ g Zn} \times \frac{1 \text{ mol}}{65.38 \text{ g}} = 0.01728 \text{ mol Zn}$$

Since the coefficients for Zn and Cl_2 in the balanced equation are the same, for 0.0173 mol of Zn reacting we will need 0.0173 mol of Cl_2

$T = 35 \,°C = 308 \text{ K}$

$$V = \frac{nRT}{P} = \frac{(0.01728 \text{ mol})(0.08206 \text{ L atm mol}^{-1} \text{ K}^{-1})(308 \text{ K})}{(1.01 \text{ atm})} = 0.433 \text{ L } Cl_2$$

90. Molar mass of $CaCO_3$ = 100.08 g

$$4.74 \text{ g } CaCO_3 \times \frac{1 \text{ mol}}{100.08 \text{ g}} = 0.0474 \text{ mol } CaCO_3$$

Since the coefficients of $CaCO_3$ and CO_2 in the balanced chemical equation are the same, when 0.0474 mol $CaCO_3$ is heated, 0.0474 mol CO_2 results.

$T = 26 \,°C = 299 \text{ K}$

$$V = \frac{nRT}{P} = \frac{(0.0474 \text{ mol})(0.08206 \text{ L atm mol}^{-1} \text{ K}^{-1})(299 \text{ K})}{(0.997 \text{ atm})} = 1.17 \text{ L } CO_2$$

92. Molar mass of Mg_3N_2 = 100.95 g

$$10.3 \text{ g } Mg_3N_2 \times \frac{1 \text{ mol}}{100.95 \text{ g}} = 0.102 \text{ mol } Mg_3N_2$$

From the balanced chemical equation, the amount of NH_3 produced will be

$$0.102 \text{ mol } Mg_3N_2 \times \frac{2 \text{ mol } NH_3}{1 \text{ mol } Mg_3N_2} = 0.204 \text{ mol } NH_3$$

$T = 24 \,°C = 297 \text{ K}$ $\qquad$ $P = 752 \text{ mm Hg} = 0.989 \text{ atm}$

$$V = \frac{nRT}{P} = \frac{(0.204 \text{ mol})(0.08206 \text{ L atm mol}^{-1} \text{ K}^{-1})(297 \text{ K})}{(0.989 \text{ atm})} = 5.03 \text{ L}$$

This assumes that the ammonia was collected dry.

Chapter 13   Gases   169

94. Molar masses: He, 4.003 g; N_2, 28.02 g

$$2.01 \text{ g He} \times \frac{1 \text{ mol}}{4.003 \text{ g}} = 0.502 \text{ mol He}$$

$$4.14 \text{ g } N_2 \times \frac{1 \text{ mol}}{28.02 \text{ g}} = 0.148 \text{ mol } N_2$$

$n_{total} = 0.502 \text{ mol} + 0.148 \text{ mol} = 0.650 \text{ mol}$

$T = 45 \text{ °C} = 318 \text{ K}$

$$V = \frac{nRT}{P} = \frac{(0.650 \text{ mol})(0.08206 \text{ L atm mol}^{-1} \text{ K}^{-1})(318 \text{ K})}{(1.02 \text{ atm})} = 16.6 \text{ L}$$

96. $P_1 = 1.47 \text{ atm}$ $P_2 = 1.00 \text{ atm}$ (Standard Pressure)
    $V_1 = 145 \text{ mL}$ $V_2 = ? \text{ mL}$
    $T_1 = 44 \text{ °C} = 317 \text{ K}$ $T_2 = 0 \text{ °C} = 273 \text{ K}$ (Standard Temperature)

$$V_2 = \frac{P_1 V_1 T_2}{P_2 T_1} = \frac{(1.47 \text{ atm})(145 \text{ mL})(273 \text{ K})}{(1.00 \text{ atm})(317 \text{ K})} = 184 \text{ mL}$$

98. Molar masses: He, 4.003 g; Ne, 20.18 g

$$6.25 \text{ g He} \times \frac{1 \text{ mol}}{4.003 \text{ g}} = 1.561 \text{ mol He}$$

$$4.97 \text{ g Ne} \times \frac{1 \text{ mol}}{20.18 \text{ g}} = 0.2463 \text{ mol Ne}$$

$n_{total} = 1.561 \text{ mol} + 0.2463 \text{ mol} = 1.807 \text{ mol}$

Since 1 mol of an ideal gas occupies 22.4 L at STP, the volume is given by

$$1.807 \text{ mol} \times \frac{22.4 \text{ L}}{1 \text{ mol}} = 40.48 \text{ L} = 40.5 \text{ L}$$

The partial pressure of a given gas in a mixture will be proportional to what *fraction* of the total number of moles of gas the given gas represents

$$P_{He} = \frac{1.561 \text{ mol He}}{1.807 \text{ mol}} \times 1.00 \text{ atm} = 0.8639 \text{ atm} = 0.864 \text{ atm}$$

$$P_{Ne} = \frac{0.2463 \text{ mol Ne}}{1.807 \text{ mol}} \times 1.00 \text{ atm} = 0.1363 \text{ atm} = 0.136 \text{ atm}$$

100. Molar mass of N_2 = 28.02 g

$$10.2 \text{ g } N_2 \times \frac{1 \text{ mol}}{28.02 \text{ g}} = 0.364 \text{ mol } N_2$$

$$0.364 \text{ mol } N_2 \times \frac{3 \text{ mol } Cl_2}{1 \text{ mol } N_2} = 1.09 \text{ mol } Cl_2$$

170   Chapter 13   Gases

$$1.09 \text{ mol Cl}_2 \times \frac{22.4 \text{ L}}{1 \text{ mol}} = 24.5 \text{ L Cl}_2$$

102. $KMnO_4(aq) + Cl_2(g) \rightarrow 2KMnO_4(s) + 2KCl(aq)$

molar mass $KMnO_4$ = 158.0 g

$$10.0 \text{ g KMnO}_4 \times \frac{1 \text{ mol KMnO}_4}{158.0 \text{ g KMnO}_4} = 0.06329 \text{ mol KMnO}_4$$

$$0.06329 \text{ mol KMnO}_4 \times \frac{1 \text{ mol Cl}_2}{2 \text{ mol KMnO}_4} = 0.03165 \text{ mol Cl}_2$$

$$0.03165 \text{ mol Cl}_2 \times \frac{22.4 \text{ L}}{1 \text{ mol}} = 0.709 \text{ L} = 709 \text{ mL}$$

104. twice

106. a. $PV = k$; $P_1V_1 = P_2V_2$
     b. $V = bT$; $V_1/T_1 = V_2/T_2$
     c. $V = an$; $V_1/n_1 = V_2/n_2$
     d. $PV = nRT$
     e. $P_1V_1/T_1 = P_2V_2/T_2$

108. First determine what volume the helium in the tank would have if it were at a pressure of 755 mm Hg (corresponding to the pressure the gas will have in the balloons).

8.40 atm = 6384 mm Hg

$$V_2 = (25.2 \text{ L}) \times \frac{6384 \text{ mm Hg}}{755 \text{ mm Hg}} = 213 \text{ L}$$

Allowing for the fact that 25.2 L of He will have to remain in the tank, this leaves 213 − 25.2 = 187.8 L of He for filling the balloons.

$$187.8 \text{ L He} \times \frac{1 \text{ balloon}}{1.50 \text{ L He}} = 125 \text{ balloons}$$

110. The volume of a gas sample is proportional to the number of moles of gas present in the sample (Avogadro's law). Therefore, the volumes of gases combining in chemical reactions should follow the stoichiometric ratios of the balanced chemical equation for the reaction, assuming the gases involved are all subject to the same conditions of temperature and pressure.

For the reaction $2NO_2(g) \rightarrow N_2O_4(g)$, we ordinarily would interpret the equation as meaning that two *moles* of NO_2 will combine to produce one *mole* of N_2O_4. If the NO_2 and N_2O_4 are each at the same pressure and temperature, however, we could just as easily interpret the equation to mean that two *volumes* of NO_2 combine to give one *volume* of N_2O_4.

$$25.0 \text{ mL NO}_2 \times \frac{1 \text{ mL N}_2\text{O}_4}{2 \text{ mL NO}_2} = 12.5 \text{ mL N}_2\text{O}_4$$

112. $CaCO_3(s) + 2H^+(aq) \rightarrow Ca^{2+}(aq) + H_2O(l) + CO_2(g)$

    molar mass $CaCO_3$ = 100.1 g

    $$10.0 \text{ g CaCO}_3 \times \frac{1 \text{ mol CaCO}_3}{100.1 \text{ g CaCO}_3} = 0.0999 \text{ mol CaCO}_3 = 0.0999 \text{ mol CO}_2 \text{ also}$$

    60°C + 273 = 333 K

    $P_{\text{carbon dioxide}} = P_{\text{total}} - P_{\text{water vapor}}$
    $P_{\text{carbon dioxide}}$ = 774 mm Hg − 149.4 mm Hg = 624.6 mm Hg = 0.822 atm

    $$V_{\text{wet}} = \frac{(0.0999 \text{ mol})(0.08206 \text{ L atm mol}^{-1} \text{ K}^{-1})(333 \text{ K})}{(0.822 \text{ atm})} = 3.32 \text{ L wet CO}_2$$

    $$V_{\text{dry}} = 3.32 \text{ L} \times \frac{624.6 \text{ mm Hg}}{774 \text{ mm Hg}} = 2.68 \text{ L}$$

114. $2KClO_3(s) \rightarrow 2KCl(s) + 3O_2(g)$

    molar mass $KClO_3$ = 122.6 g

    $$50.0 \text{ g KClO}_3 \times \frac{1 \text{ mol KClO}_3}{122.6 \text{ g KClO}_3} = 0.408 \text{ mol KClO}_3$$

    $$0.408 \text{ mol KClO}_3 \times \frac{3 \text{ mol O}_2}{2 \text{ mol KClO}_3} = 0.612 \text{ mol O}_2$$

    25°C + 273 = 298 K   630. torr = 0.829 atm

    $$V = nRT/P = \frac{(0.612)(0.08206 \text{ L atm mol}^{-1} \text{ K}^{-1})(298 \text{ K})}{(0.829 \text{ atm})} = 18.1 \text{ L O}_2$$

116. a. $665 \text{ mm Hg} \times \dfrac{1 \text{ atm}}{760 \text{ mm Hg}} = 0.875 \text{ atm}$

    b. $124 \text{ kPa} = 124 \times 10^3 \text{ Pa}$

    $124 \times 10^3 \text{ Pa} \times \dfrac{1 \text{ atm}}{101{,}325 \text{ Pa}} = 1.22 \text{ atm}$

    c. $2.540 \times 10^6 \text{ Pa} \times \dfrac{1 \text{ atm}}{101{,}325 \text{ Pa}} = 25.07 \text{ atm}$

    d. $803 \text{ torr} \times \dfrac{1 \text{ atm}}{760 \text{ torr}} = 1.06 \text{ atm}$

118. a. $645 \text{ mm Hg} \times \dfrac{1 \text{ atm}}{760 \text{ mm Hg}} \times \dfrac{101{,}325 \text{ Pa}}{1 \text{ atm}} = 8.60 \times 10^4 \text{ Pa}$

    b. $221 \text{ kPa} = 221 \times 10^3 \text{ Pa} = 2.21 \times 10^5 \text{ Pa}$

172   Chapter 13   Gases

    c. $0.876 \text{ atm} \times \dfrac{101,325 \text{ Pa}}{1 \text{ atm}} = 8.87 \times 10^4 \text{ Pa}$

    d. $32 \text{ torr} \times \dfrac{1 \text{ atm}}{760 \text{ torr}} \times \dfrac{101,325 \text{ Pa}}{1 \text{ atm}} = 4.3 \times 10^3 \text{ Pa}$

120. a.   1.00 mm Hg = 1.00 torr

           $V = 255 \text{ mL} \times \dfrac{1.00 \text{ torr}}{2.00 \text{ torr}} = 128 \text{ mL}$

    b.   1.0 atm = 101.325 kPa

           $V = 1.3 \text{ L} \times \dfrac{1.0 \text{ kPa}}{101.325 \text{ kPa}} = 1.3 \times 10^{-2} \text{ L}$

    c.   1.0 mm Hg = 0.133 kPa

           $V = 1.3 \text{ L} \times \dfrac{1.0 \text{ kPa}}{0.133 \text{ kPa}} = 9.8 \text{ L}$

122. 1.105 L = 1105 mL

      $755 \text{ torr} \times \dfrac{1105 \text{ mL}}{1.00 \text{ mL}} = 8.34 \times 10^5 \text{ torr} \ (1.10 \times 10^3 \text{ atm})$

124. a.   75°C + 273 = 348 K   -75°C + 273 = 198 K

           $100. \text{ mL} \times \dfrac{198 \text{ K}}{348 \text{ K}} = 56.9 \text{ mL}$

    b.   100°C + 273 = 373 K

           $373 \text{ K} \times \dfrac{600 \text{ mL}}{500 \text{ mL}} = 448 \text{ K} \ (175°C)$

    c.   zero (the volume of any gas sample becomes zero at 0 K)

126. 25°C + 273 = 298 K   -100°C + 273 = 173 K

    $25 \text{ L} \times \dfrac{173 \text{ K}}{298 \text{ K}} = 15 \text{ L}$

128. For a given gas, the number of moles present in a sample is directly proportional to the mass of the sample. So the problem can be solved even though the gas is not identified (so that its molar mass is not known).

    $23.2 \text{ g} \times \dfrac{10.4 \text{ L}}{93.2 \text{ L}} = 2.59 \text{ g}$

130. a.   V = 21.2 mL = 0.0212 L

        $T = PV/nR = \dfrac{(1.034 \text{ atm})(0.0212 \text{ L})}{(0.00432 \text{ mol})(0.08206 \text{ L atm mol}^{-1} \text{ K}^{-1})} = 61.8 \text{ K}$

Chapter 13   Gases   173

b. $V = 1.73$ mL $= 0.00173$ L

$$P = nRT/V = \frac{(0.000115 \text{ mol})(0.08206 \text{ L atm mol}^{-1} \text{ K}^{-1})(182 \text{ K})}{(0.00173 \text{ L})}$$

$P = 0.993$ atm

c. $P = 1.23$ mm Hg $= 0.00162$ atm

$T = 152°C + 273 = 425$ K

$$V = nRT/P = \frac{(0.773 \text{ mol})(0.08206 \text{ L atm mol}^{-1} \text{ K}^{-1})(425 \text{ K})}{(0.00162 \text{ atm})}$$

$V = 1.67 \times 10^4$ L

132. $27°C + 273 = 300$ K

The number of moles of gas it takes to fill the 100. L tanks to 120 atm at 27°C is independent of the identity of the gas.

$$n = PV/RT = \frac{(120 \text{ atm})(100. \text{ L})}{(0.08206 \text{ L atm mol}^{-1} \text{ K}^{-1})(300 \text{ K})} = 487 \text{ mol}$$

487 mol of *any* gas will fill the tanks to the required specifications.

molar masses: CH_4, 16.0 g; N_2, 28.0 g; CO_2, 44.0 g

for CH_4: (487 mol)(16.0 g/mol) = 7792 g = 7.79 kg CH_4

for N_2: (487 mol)(28.0 g/mol) = 13,636 g = 13.6 kg N_2

for CO_2: (487 mol)(44.0 g/mol) = 21,428 g = 21.4 kg CO_2

134. molar mass of O_2 = 32.00 g

55 mg = 0.055 g

$$n = 0.055 \text{ g} \times \frac{1 \text{ mol}}{32.00 \text{ g}} = 0.0017 \text{ mol}$$

$V = 100.$ mL $= 0.100$ L

$T = 26°C + 273 = 299$ K

$$P = nRT/V = \frac{(0.0017 \text{ mol})(0.08206 \text{ L atm mol}^{-1} \text{ K}^{-1})(299 \text{ K})}{(0.100 \text{ L})} = 0.42 \text{ atm}$$

136. $P_1 = 1.13$ atm $P_2 = 1.89$ atm

$V_1 = 100$ mL $= 0.100$ L $V_2 = 500$ mL $= 0.500$ L

$T_1 = 300$ K $T_2 = ?$

$$T_2 = \frac{T_1 P_2 V_2}{P_1 V_1} = \frac{(300 \text{ K})(1.89 \text{ atm})(0.500 \text{ L})}{(1.13 \text{ atm})(0.100 \text{ L})} = 2.51 \times 10^3 \text{ K}$$

Note that the calculation could have been carried through with the two volumes expressed in milliliters since the universal gas constant does not appear explicitly in this form of the ideal gas equation.

138. molar masses: O_2, 32.0 g; N_2, 28.0 g; CO_2, 44.0 g; Ne, 20.2 g

$$5.0 \text{ g } O_2 \times \frac{1 \text{ mol } O_2}{32.0 \text{ g } O_2} = 0.156 \text{ mol } O_2$$

$$5.0 \text{ g } N_2 \times \frac{1 \text{ mol } N_2}{28.0 \text{ g } N_2} = 0.179 \text{ mol } N_2$$

$$5.0 \text{ g } CO_2 \times \frac{1 \text{ mol } CO_2}{44.0 \text{ g } CO_2} = 0.114 \text{ mol } CO_2$$

$$5.0 \text{ g Ne} \times \frac{1 \text{ mol Ne}}{20.2 \text{ g Ne}} = 0.248 \text{ mol Ne}$$

Total moles of gas = 0.156 + 0.179 + 0.114 + 0.248 = 0.697 mol

29°C + 273 = 302 K

755 mm Hg = 0.993 atm

$$V = nRT/P = \frac{(0.697 \text{ mol})(0.08206 \text{ L atm mol}^{-1} \text{ K}^{-1})(302 \text{ K})}{(0.993 \text{ atm})} = 17 \text{ L}$$

140. molar mass of NH_3 = 17.03 g

$$5.00 \text{ g } NH_3 \times \frac{1 \text{ mol } NH_3}{17.03 \text{ g } NH_3} = 0.294 \text{ mol } NH_3 \text{ to be produced}$$

$N_2(g) + 3H_2(g) \rightarrow 2NH_3(g)$

$$0.294 \text{ mol } NH_3 \times \frac{1 \text{ mol } N_2}{2 \text{ mol } NH_3} = 0.147 \text{ mol } N_2 \text{ required}$$

$$0.294 \text{ mol } NH_3 \times \frac{3 \text{ mol } H_2}{2 \text{ mol } NH_3} = 0.441 \text{ mol } H_2 \text{ required}$$

11°C + 273 = 284 K

$$V_{nitrogen} = \frac{(0.147 \text{ mol})(0.08206 \text{ L atm mol}^{-1} \text{ K}^{-1})(284 \text{ K})}{(0.998 \text{ atm})} = 3.43 \text{ L } N_2$$

$$V_{hydrogen} = \frac{(0.441 \text{ mol})(0.08206 \text{ L atm mol}^{-1} \text{ K}^{-1})(284 \text{ K})}{(0.998 \text{ atm})} = 10.3 \text{ L } H_2$$

142. $2Cu_2S(s) + 3O_2(g) \rightarrow 2Cu_2O(s) + 2SO_2(g)$

molar mass Cu_2S = 159.2 g

$$25 \text{ g } Cu_2S \times \frac{1 \text{ mol } Cu_2S}{159.2 \text{ g } Cu_2S} = 0.1570 \text{ mol } Cu_2S$$

$$0.1570 \text{ mol } Cu_2S \times \frac{3 \text{ mol } O_2}{2 \text{ mol } Cu_2S} = 0.2355 \text{ mol } O_2$$

27.5°C + 273 = 300.5 K

$$V_{oxygen} = \frac{(0.2355 \text{ mol})(0.08206 \text{ L atm mol}^{-1} \text{ K}^{-1})(300.5 \text{ K})}{(0.998 \text{ atm})} = 5.8 \text{ L O}_2$$

$$0.1570 \text{ mol Cu}_2\text{S} \times \frac{2 \text{ mol SO}_2}{2 \text{ mol Cu}_2\text{S}} = 0.1570 \text{ mol SO}_2$$

$$V_{sulfur\ dioxide} = \frac{(0.1570 \text{ mol})(0.08206 \text{ L atm mol}^{-1} \text{ K}^{-1})(300.5 \text{ K})}{(0.998 \text{ atm})}$$

$$V_{sulfur\ dioxide} = 3.9 \text{ L SO}_2$$

144. One mole of any ideal gas occupies 22.4 L at STP.

$$35 \text{ mol N}_2 \times \frac{22.4 \text{ L}}{1 \text{ mol}} = 7.8 \times 10^2 \text{ L}$$

146. molar masses: He, 4.003 g; Ar, 39.95 g; Ne, 20.18 g

$$5.0 \text{ g He} \times \frac{1 \text{ mol He}}{4.003 \text{ g He}} = 1.249 \text{ mol He}$$

$$1.0 \text{ g Ar} \times \frac{1 \text{ mol Ar}}{39.95 \text{ g Ar}} = 0.02503 \text{ mol Ar}$$

$$3.5 \text{ g Ne} \times \frac{1 \text{ mol Ne}}{20.18 \text{ g Ne}} = 0.1734 \text{ mol Ne}$$

Total moles of gas = 1.249 + 0.02503 + 0.1734 = 1.447 mol

22.4 L is the volume occupied by one mole of any ideal gas at STP. This would apply even if the gas sample is a *mixture* of individual gases.

$$1.447 \text{ mol} \times \frac{22.4 \text{ L}}{1 \text{ mol}} = 32 \text{ L total volume for the mixture}$$

The *partial pressure* of each individual gas in the mixture will be related to what *fraction* on a mole basis each gas represents in the mixture.

$$P_{He} = 1.00 \text{ atm} \times \frac{1.249 \text{ mol He}}{1.447 \text{ mol total}} = 0.86 \text{ atm}$$

$$P_{Ar} = 1.00 \text{ atm} \times \frac{0.02503 \text{ mol Ar}}{1.447 \text{ mol total}} = 0.017 \text{ atm}$$

$$P_{Ne} = 1.00 \text{ atm} \times \frac{0.1734 \text{ mol Ne}}{1.447 \text{ mol total}} = 0.12 \text{ atm}$$

148. The solution is only 50% H_2O_2. Therefore 125 g solution = 62.5 g H_2O_2

Molar mass of H_2O_2 = 34.02 g

$$62.5 \text{ g H}_2\text{O}_2 \times \frac{1 \text{ mol}}{34.02 \text{ g}} = 1.84 \text{ mol H}_2\text{O}_2$$

$$1.84 \text{ mol } H_2O_2 \times \frac{1 \text{ mol } O_2}{2 \text{ mol } H_2O_2} = 0.920 \text{ mol } O_2$$

$T = 27\ °C = 300\ K \qquad\qquad P = 764 \text{ mm Hg} = 1.01 \text{ atm}$

$$V = nRT/P = \frac{(0.920)(0.08206 \text{ L atm mol}^{-1} \text{ K}^{-1})(300 \text{ K})}{(1.01 \text{ atm})} = 22.5 \text{ L}$$

Chapter 14 Liquids and Solids

2. Water exerts its cooling effect in nature in many ways. Water, as perspiration, helps cool the human body (the evaporation of water from skin is an endothermic process; the heat required for evaporation comes from the body). Large bodies of natural water (e.g., the oceans) have a cooling effect on nearby land masses (the interior of the United States, away from the oceans, tends to be hotter than coastal regions). In industry, water is used as a coolant in *many* situations Some nuclear power plants, for example, use water to cool the reactor core. Many office buildings are air-conditioned in summer by circulating cold water systems.

4. The fact that water expands when it freezes often results in broken water pipes during cold weather. The expansion of water when it freezes also makes ice float on liquid water. The expansion of a given mass of water into a larger volume upon freezing lowers the density of ice compared to liquid water. Aquatic life (and probably all life) could not exist if ice sank in water.

6. Sloped portions of a heating/cooling curve represent *changes in temperature* as heat is applied or removed; for example, in the cooling/heating curve shown, there are sloped portions representing the heating of ice, the heating of liquid water, and the heating of steam, as heat continues to be applied. Flat portions of such curves represent equilibrium transitions between states; for example, the flat portions in the curve shown represent the solid-liquid (melting-freezing) transition and the liquid-vapor (boiling-condensation) phase transitions.

8. As a sample of liquid water is heated, the motions of the molecules increase as the temperature rises. As the liquid reaches its boiling point, bubbles of vapor begin to form in the liquid, which rise to the surface of the liquid and burst. As the liquid remains at its boiling point, the additional heat energy being supplied to the liquid is used to overcome attractive forces among the molecules in the liquid. As heat energy continues to be applied, more and more molecules will be moving in the right direction and with sufficient energy to escape.

10. The additional energy is used to overcome attractive forces among the molecules in the liquid. As molecules move faster, they spend less time near each other, and there are smaller intermolecular forces built up.

12. In a vapor, the molecules are so far apart that intermolecular forces become negligible, whereas in a liquid, the intermolecular forces are still quite strong. It takes more heat to vaporize a liquid than to melt the same amount of solid because of the greater degree to which the intermolecular forces must be overcome.

14. $10.0 \text{ g X} \times \dfrac{1 \text{ mol X}}{52 \text{ g X}} = 0.192 \text{ mol X}$

   $0.192 \text{ mol} \times \dfrac{2.5 \text{ kJ}}{1 \text{ mol}} = 0.48 \text{ kJ}$

178   Chapter 14   Liquids and Solids

$$25.0 \text{ g X} \times \frac{1 \text{ mol X}}{52 \text{ g X}} = 0.481 \text{ mol X}$$

$$0.481 \text{ mol} \times \frac{55.3 \text{ kJ}}{1 \text{ mol}} = 27 \text{ kJ}$$

16. molar mass H_2O = 18.0 g

$$5.0 \text{ g } H_2O \times \frac{1 \text{ mol } H_2O}{18.0 \text{ g } H_2O} = 0.278 \text{ mol } H_2O$$

To melt the ice: $0.278 \text{ mol} \times \dfrac{6.02 \text{ kJ}}{1 \text{ mol}} = 1.67 \text{ kJ}$

To heat the liquid: $5.0 \text{ g} \times 4.18 \dfrac{J}{g \cdot °C} \times 100°C = 2090 \text{ J} = 2.09 \text{ kJ}$

To vaporize the liquid: $0.278 \text{ mol} \times \dfrac{40.6 \text{ kJ}}{1 \text{ mol}} = 11.27 \text{ kJ}$

Total heat required = 1.67 + 2.09 + 11.27 = 15.03 kJ = 15 kJ

18. The *molar* heat of fusion of aluminum is the heat required to melt 1 mol.

$$\frac{3.95 \text{ kJ}}{1 \text{ g}} \times \frac{26.98 \text{ g}}{1 \text{ mol}} = 107 \text{ kJ/mol}$$

$$10.0 \text{ g} \times \frac{3.95 \text{ kJ}}{1 \text{ g}} = 39.5 \text{ kJ required to melt 10.0 g Al}$$

$$10.0 \text{ mol} \times \frac{107 \text{ kJ}}{1 \text{ mol}} = 1.07 \times 10^3 \text{ kJ required to melt 10.0 mol Al.}$$

20. Dipole-dipole forces are relatively stronger at short distances; they are short-range forces. Molecules must first closely approach one another before dipole-dipole forces can cause attraction between molecules.

22. Water molecules are able to form strong *hydrogen bonds* with each other. These bonds are an especially strong form of dipole-dipole forces and are only possible when hydrogen atoms are bonded to the most electronegative elements (N, O, and F). The extra strong intermolecular forces in H_2O require much higher temperatures (high energies) to be overcome to permit the liquid to boil. We take the fact that water has a much higher boiling point that the other hydrogen compounds of the Group 6 elements as proof that a special force is at play in water (hydrogen bonding).

24. The fact that such nonpolar, monatomic molecules *can* be liquefied and solidified indicates that there must be *some* sort of intermolecular forces possible between atoms in these substances. London dispersion forces arise when a temporary (instantaneous) dipolar arrangement of

charge develops as the electrons of an atom move around its nucleus. The positive nucleus of one atom can then weakly attract the displaced electron cloud of another atom.

26. a. London dispersion forces (nonpolar, monatomic molecules)

    b. London dispersion forces (nonpolar molecules)

    c. dipole-dipole forces; London dispersion forces

    d. hydrogen bonding (H attached to O); London dispersion forces

28. An increase in the heat of fusion is observed for an increase in the size of the halogen atom involved (the electron cloud of a larger atom is more easily polarized by an approaching dipole, thus giving larger London dispersion forces).

30. For a homogeneous mixture to be able to form at all, the forces between molecules of the two substances being mixed must be at least *comparable in magnitude* to the intermolecular forces within each *separate* substance. Apparently in the case of a water-ethanol mixture, the forces that exist when water and ethanol are mixed are stronger than water-water or ethanol-ethanol forces in the separate substances. This allows ethanol and water molecules to approach each other more closely in the mixture than either substance's molecules could approach a like molecule in the separate substances. There is strong hydrogen bonding in both ethanol and water.

32. Vapor pressure is the pressure of vapor present *at equilibrium* above a liquid in a sealed container at a particular temperature. When a liquid is placed in a closed container, molecules of the liquid evaporate freely into the empty space above the liquid. As the number of molecules present in the vapor state increases with time, vapor molecules begin to rejoin the liquid state (condense). Eventually a dynamic equilibrium is reached between evaporation and condensation in which the net number of molecules present in the vapor phase becomes *constant* with time.

34. A method is shown in Figure 14.10. The apparatus consists of basically a barometer into which a volatile liquid may be injected. Since mercury is so much more dense than other liquids, the injected volatile liquid rises to the top of the mercury column in the tube and floats on top of the mercury. Since the space above the mercury is a vacuum, the volatile liquid evaporates into the empty space. As the liquid is converted to the gaseous state, the level of the mercury column drops as the pressure of vapor builds up.

36. a. HF   Although both substances are capable of hydrogen bonding, water has two O—H bonds which can be involved in hydrogen bonding versus only one F—H bond in HF.

    b. CH_3OCH_3 Since there is no H attached to the O atom, no hydrogen bonding can exist. Since there is no hydrogen bonding possible, the molecule should be relatively more volatile than CH_3CH_2OH even though it contains the same number of atoms of each element.

    c. CH_3SH Hydrogen bonding is not as important for a S–H bond (because S has a lower electronegativity than O). Since there is little hydrogen bonding, the molecule is relatively more volatile than CH_3OH.

38. Both substances have the same molar mass. However ethyl alcohol contains a hydrogen atom directly bonded to an oxygen atom. Therefore, hydrogen bonding can exist in ethyl alcohol, whereas only weak dipole-dipole forces can exist in dimethyl ether. Dimethyl ether is more volatile; ethyl alcohol has a higher boiling point.

40. *Ionic* solids have as their fundamental particles positive and negative *ions*; a simple example is sodium chloride, in which Na^+ and Cl^- ions are held together by strong electrostatic forces.

    Molecular solids have molecules as their fundamental particles, with the molecules being held together in the crystal by dipole-dipole forces, hydrogen bonding forces, or London dispersion forces (depending on the identity of the substance); simple examples of molecular solids include ice (H_2O) and ordinary table sugar (sucrose).

    Atomic solids have simple atoms as their fundamental particles, with the atoms being held together in the crystal either by covalent bonding (as in graphite or diamond) or by metallic bonding (as in copper or other metals).

42. The fundamental particles in molecular solids are individual *molecules*. Examples are *ice* (individual water molecules), *table sugar* (individual sucrose molecules), *dry ice* (individual carbon dioxide molecules), etc.

44. Ionic solids consist of a crystal lattice of basically alternating positively and negatively charged ions. A given ion is surrounded by several ions of the opposite charge, all of which electrostatically attract it strongly. This pattern repeats itself throughout the crystal. Sometimes a crystal of an ionic substance is described as being like one giant molecule, with all the ions bonded to each other.

46. Ordinary ice contains nonlinear, highly polar water molecules. In addition, there is extensive, strong hydrogen bonding possible between water molecules in ordinary ice. Dry ice, on the other hand, consists of linear, nonpolar carbon dioxide molecules, and only very weak intermolecular forces are possible.

48. In a metal, the valence electrons are mobile, and can move throughout the entire metal crystal's lattice. In ionic solids, although there are positive and negative ions present, each ion is held rigidly in place by several ions of the opposite charge.

50. Alloys may be of two types: *substitutional* (in which one metal is substituted for another in the regular positions of the crystal lattice) and *interstitial* (in which a second metal's atoms fit into the empty space in a given metal's crystal lattice). The presence of atoms of a second metal in a given metal's crystal lattice changes the properties of the metal: frequently the alloy is stronger than either of the original metals because the irregularities introduced into the crystal lattice by the presence of a second metal's atoms prevent the crystal from being deformed as easily. The properties of iron may be modified by alloying with many different substances, particularly with carbon, nickel, and cobalt. Steels with relatively high carbon content are exceptionally strong, whereas steels with low carbon contents are softer, more malleable, and more ductile. Steels produced by alloying iron with nickel and cobalt are more resistant to corrosion than iron itself.

52. j

54. f

56. d

58. a

60. l

62. Dimethyl ether has the larger vapor pressure. No hydrogen bonding is possible since the O atom does not have a hydrogen atom attached. Hydrogen bonding can occur *only* when a hydrogen atom is *directly* attached to a strongly electronegative atom (such as N, O, or F). Hydrogen bonding *is* possible in ethanol (ethanol contains an –OH group).

64. a. H_2. London dispersion forces are the only intermolecular forces present in these nonpolar molecules; typically London forces become larger with increasing atomic size (as the atoms become bigger, the edge of the electron cloud lies farther from the nucleus and becomes more easily distorted).

    b. Xe. Only the relatively weak London forces could exist in a crystal of Xe atoms, whereas in NaCl strong ionic forces exist, and in diamond strong covalent bonding exists between carbon atoms.

    c. Cl_2. Only London forces exist among such nonpolar molecules. London forces become larger with increasing atomic size.

66. Steel is a general term applied to alloys consisting primarily of iron, but with small amounts of other substances added. Whereas pure iron itself is relatively soft, malleable, and ductile, steels are typically much stronger and harder, and much less subject to damage.

68. Water is the solvent in which cellular processes take place in living creatures. Water in the oceans moderates the earth's temperature. Water is used in industry as a cooling agent. Water serves as a means of transportation on the earth's oceans. The liquid range is 0°C to 100°C at 1 atm pressure.

70. At higher altitudes, the boiling points of liquids, such as water, are lower because there is a lower atmospheric pressure above the liquid. The temperature at which food cooks is determined by the temperature to which the water in the food can be heated before it escapes as steam. Thus, food cooks at a lower temperature at high elevations where the boiling point of water is lowered.

72. Heat of fusion (melt); Heat of vaporization (boil).
The heat of vaporization is always larger, because virtually all of the intermolecular forces must be overcome to form a gas. In a liquid, considerable intermolecular forces remain. Thus going from a solid to liquid requires less energy than going from the liquid to the gas.

74. Dipole-dipole interactions are typically about 1% as strong as a covalent bond. Dipole-dipole interactions represent electrostatic attractions between portions of molecules which carry only a *partial* positive or negative charge, and such forces require the molecules that are interacting to come *near* enough to each other.

76. London dispersion forces are relatively weak forces that arise among noble gas atoms and in nonpolar molecules. London forces are due to *instantaneous dipoles* that develop when one atom (or molecule) momentarily distorts the electron cloud of another atom (or molecule). London forces are typically weaker than either permanent dipole-dipole forces or covalent bonds.

78. For every mole of liquid water that evaporates, several kiloJoules of heat must be absorbed to provide kinetic energy for the molecules. This heat is absorbed by the water from its surroundings.

80. In NH_3, strong hydrogen bonding can exist. In CH_4, because the molecule is nonpolar, only the relatively weak London dispersion forces exist.

82. In a crystal of ice, strong *hydrogen bonding* forces are present, while in the crystal of a nonpolar substance like oxygen, only the much weaker *London* forces exist.

Chapter 15 Solutions

2. A heterogeneous mixture does not have a uniform composition: the composition varies in different places within the mixture. Examples of nonhomogeneous mixtures include salad dressing (mixture of oil, vinegar, water, herbs and spices) and granite (combination of minerals).

4. solid

6. One substance will mix with and dissolve in another substance if the intermolecular forces are similar in the two substances, so that when the mixture forms, the forces between particles in the mixture will be similar to the forces present in the separate substances. Sugar and ethyl alcohol molecules both contain polar -OH groups, which are comparable to the polar -OH structure in water. Sugar or ethyl alcohol molecules can hydrogen bond with water molecules and intermingle with them freely to form a solution. Substances like petroleum (whose molecules contain only carbon and hydrogen) are very nonpolar and cannot form interactions with polar water molecules.

8. independently

10. unsaturated

12. large

14. nine

16. a. $\dfrac{1.0 \text{ g NH}_4\text{NO}_3}{(10. \text{ g H}_2\text{O} + 1.0 \text{ g NH}_4\text{NO}_3)} \times 100 = 9.1\% \text{ NH}_4\text{NO}_3$

    b. $\dfrac{1.0 \text{ g NH}_4\text{NO}_3}{(25 \text{ g H}_2\text{O} + 1.0 \text{ g NH}_4\text{NO}_3)} \times 100 = 3.8\% \text{ NH}_4\text{NO}_3$

    c. $\dfrac{1.0 \text{ g NH}_4\text{NO}_3}{(75 \text{ g H}_2) + 1.0 \text{ g NH}_4\text{NO}_3} \times 100 = 1.3\% \text{ NH}_4\text{NO}_3$

    d. $\dfrac{1.0 \text{ g NH}_4\text{NO}_3}{(100 \text{ g H}_2\text{O} + 1.0 \text{ g NH}_4\text{NO}_3)} \times 100 = 0.99\% \text{ NH}_4\text{NO}_3$

18. To say a solution is 15.0% NaCl by mass means that 100. g of the solution would contain 15.0 g of NaCl.

    a. $150. \text{ g solution} \times \dfrac{15.0 \text{ g NaCl}}{100. \text{ g solution}} = 22.5 \text{ g NaCl}$

    b. $35.0 \text{ g NaCl} \times \dfrac{100. \text{ g solution}}{15.0 \text{ g NaCl}} = 233 \text{ g solution}$

    c. $1000. \text{ g solution} \times \dfrac{15.0 \text{ g NaCl}}{100. \text{ g solution}} = 150. \text{ g NaCl}$

184 Chapter 15 Solutions

d. $\quad 1000.\text{ g solution} \times \dfrac{15.0\text{ g NaCl}}{100.\text{ g solution}} = 150.\text{ g NaCl}$

Note that the calculation for parts (c) and (d) are identical. These questions are just different ways of asking the same thing: How many g of NaCl are in 1000. g of a 15.0% NaCl solution?

20. $\%\text{Cu} = \dfrac{5.31\text{ g Cu}}{(5.31\text{ g Cu} + 4.03\text{ g Zn} + 145\text{ g Fe})} \times 100 = \dfrac{5.31\text{ g}}{154.3\text{ g}} \times 100 = 3.44\% \text{ Cu}$

$\%\text{ Zn} = \dfrac{4.03\text{ g Zn}}{(5.31\text{ g Cu} + 4.03\text{ g Zn} + 145\text{ g Fe})} \times 100 = \dfrac{4.03\text{ g}}{154.3\text{ g}} \times 100 = 2.61\% \text{ Zn}$

$\%\text{ Fe} = \dfrac{145\text{ g Fe}}{(5.31\text{ g Cu} + 4.03\text{ g Zn} + 145\text{ g Fe})} \times 100 = \dfrac{145\text{ g}}{154.3\text{ g}} \times 100 = 93.9\% \text{ Fe}$

22. $\dfrac{5.35\text{ g NaCl}}{151\text{ g solution}} \times 100 = 3.54\% \text{ NaCl}$

24. To say that the solution is 6.25% KBr by mass, means that 100. g of the solution will contain 6.25 g KBr.

$125\text{ g solution} \times \dfrac{6.25\text{ g KBr}}{100.\text{ g solution}} = 7.81\text{ g KBr}$

26. To say that the solution is to be 1.25% $CuCl_2$ by mass, means that 100. g of the solution will contain 1.25 g $CuCl_2$

$1250.\text{ g solution} \times \dfrac{1.25\text{ g } CuCl_2}{100.\text{ g solution}} = 15.6\text{ g } CuCl_2$

28. $\text{g heptane} = 93\text{ g solution} \times \dfrac{5.2\text{ g heptane}}{100.\text{ g solution}} = 4.8\text{ g heptane}$

$\text{g pentane} = 93\text{ g solution} \times \dfrac{2.9\text{ g pentane}}{100.\text{ g solution}} = 2.7\text{ g pentane}$

g hexane = 93 g solution − 4.8 g heptane − 2.7 g pentane = 86 g hexane

30. 0.50; 1.0

32. To say that a solution has a concentration of 5 M means that in 1 L of solution (not solvent) there would be 5 mol of solute; to prepare such a solution one would place 5 mol of NaCl in a 1 L flask, and then add whatever amount of water is necessary so that the *total* volume would be 1 L after mixing. The NaCl will occupy some space, so the amount of water to be added will be *less* than 1.00 L.

34. $\text{Molarity} = \dfrac{\text{moles of solute}}{\text{liters of solution}}$

a. 250 mL = 0.25 L

$$M = \frac{0.50 \text{ mol KBr}}{0.25 \text{ L solution}} = 2.0 \text{ }M$$

b. 500 mL = 0.50 L

$$M = \frac{0.50 \text{ mol KBr}}{0.50 \text{ L solution}} = 1.0 \text{ }M$$

c. 750 mL = 0.75 L

$$M = \frac{0.50 \text{ mol KBr}}{0.75 \text{ L solution}} = 0.67 \text{ }M$$

d. $M = \dfrac{0.50 \text{ mol KBr}}{1.0 \text{ L solution}} = 0.50 \text{ }M$

36. Molarity = $\dfrac{\text{moles of solute}}{\text{liters of solution}}$

a. Molar mass of $CuCl_2$ = 134.45 g   125 mL = 0.125 L

$$4.25 \text{ g } CuCl_2 \times \frac{1 \text{ mol}}{134.45 \text{ g}} = 0.0316 \text{ mol } CuCl_2$$

$$M = \frac{0.0316 \text{ mol } CuCl_2}{0.125 \text{ L solution}} = 0.253 \text{ }M$$

b. Molar mass of $NaHCO_3$ = 84.01 g   11.3 mL = 0.0113 L

$$0.101 \text{ g } NaHCO_3 \times \frac{1 \text{ mol}}{84.01 \text{ g}} = 0.00120 \text{ mol } NaHCO_3$$

$$M = \frac{0.00120 \text{ mol } NaHCO_3}{0.0113 \text{ L solution}} = 0.106 \text{ }M$$

c. Molar mass of Na_2CO_3 = 105.99 g

$$52.9 \text{ g } Na_2CO_3 \times \frac{1 \text{ mol}}{105.99 \text{ g}} = 0.499 \text{ mol } Na_2CO_3$$

$$M = \frac{0.499 \text{ mol } Na_2CO_3}{1.15 \text{ L solution}} = 0.434 \text{ }M$$

d. Molar mass of KOH = 56.11 g   1.5 mL = 0.0015 L

$$0.14 \text{ mg KOH} \times \frac{1 \text{ g}}{10^3 \text{ mg}} \times \frac{1 \text{ mol}}{56.11 \text{ g}} = 2.50 \times 10^{-6} \text{ mol KOH}$$

$$M = \frac{2.50 \times 10^{-6} \text{ mol KOH}}{0.0015 \text{ L solution}} = 1.67 \times 10^{-3} \text{ }M = 1.7 \times 10^{-3} \text{ }M$$

186   Chapter 15   Solutions

38. Molar mass of $CuSO_4$ = 159.6 g

$$275 \text{ g } CuSO_4 \times \frac{1 \text{ mol}}{159.6 \text{ g}} = 1.72 \text{ mol } CuSO_4$$

$$M = \frac{1.72 \text{ mol } CuSO_4}{4.25 \text{ L solution}} = 0.405 \text{ } M$$

40. Molar mass of I_2 = 253.8 g

225 mL = 0.225 L

$$5.15 \text{ g } I_2 \times \frac{1 \text{ mol}}{253.8 \text{ g}} = 0.0203 \text{ mol } I_2$$

$$M = \frac{0.0203 \text{ mol } I_2}{0.225 \text{ L solution}} = 0.0902 \text{ } M$$

42. Molar mass of NaOH = 40.00 g

$$495 \text{ g NaOH} \times \frac{1 \text{ mol}}{40.00 \text{ g}} = 12.4 \text{ mol NaOH}$$

$$M = \frac{12.4 \text{ mol NaOH}}{20.0 \text{ L solution}} = 0.619 \text{ } M$$

44. a. molar mass of HCl = 36.46 g

$$1.29 \text{ L solution} \times \frac{0.411 \text{ mol HCl}}{1.00 \text{ L solution}} = 0.530 \text{ mol HCl}$$

$$0.530 \text{ mol HCl} \times \frac{36.46 \text{ g HCl}}{1 \text{ mol HCl}} = 19.3 \text{ g HCl}$$

  b. molar mass of $CaCl_2$ = 111.0 g   145 mL = 0.145 L

$$0.145 \text{ L solution} \times \frac{0.215 \text{ mol } CaCl_2}{1.00 \text{ L solution}} = 0.0312 \text{ mol } CaCl_2$$

$$0.0312 \text{ mol } CaCl_2 \times \frac{111.0 \text{ g } CaCl_2}{1 \text{ mol } CaCl_2} = 3.46 \text{ g } CaCl_2$$

  c. molar mass HNO_3 = 63.02 g   24.9 mL = 0.0249 L

$$0.0249 \text{ L solution} \times \frac{1.15 \text{ mol } HNO_3}{1.00 \text{ L solution}} = 0.0286 \text{ mol } HNO_3$$

$$0.0286 \text{ mol } HNO_3 \times \frac{63.02 \text{ g } HNO_3}{1 \text{ mol } HNO_3} = 1.80 \text{ g } HNO_3$$

  d. molar mass of NaCl = 58.44 g

$$1.01 \text{ L solution} \times \frac{5.41 \text{ mol NaCl}}{1.00 \text{ L solution}} = 5.46 \text{ mol NaCl}$$

$$5.46 \text{ mol NaCl} \times \frac{58.44 \text{ g NaCl}}{1 \text{ mol NaCl}} = 319 \text{ g NaCl}$$

46. a. 145 mL = 0.145 L

   $$0.145 \text{ L solution} \times \frac{0.0221 \text{ mol EDTA}}{1.00 \text{ L solution}} = 3.205 \times 10^{-3} \text{ mol EDTA}$$

   Molar mass of EDTA = 292.2 g

   $$3.205 \times 10^{-3} \text{ mol EDTA} \times \frac{292.2 \text{ g}}{1 \text{ mol}} = 0.936 \text{ g EDTA}$$

   b. $$2.25 \text{ L solution} \times \frac{0.135 \text{ mol CuSO}_4}{1.00 \text{ L solution}} = 0.304 \text{ mol CuSO}_4$$

   Molar mass of $CuSO_4$ = 159.6 g

   $$0.304 \text{ mol CuSO}_4 \times \frac{159.6 \text{ g}}{1 \text{ mol}} = 48.5 \text{ g CuSO}_4$$

   c. 1.02 mL = 0.00102 L

   $$0.00102 \text{ L solution} \times \frac{0.452 \text{ mol H}_3\text{PO}_4}{1.00 \text{ L solution}} = 4.61 \times 10^{-4} \text{ mol H}_3\text{PO}_4$$

   Molar mass of H_3PO_4 = 97.99 g

   $$4.61 \times 10^{-4} \text{ mol H}_3\text{PO}_4 \times \frac{97.99 \text{ g}}{1 \text{ mol}} = 0.0452 \text{ g H}_3\text{PO}_4$$

   d. 24 mL = 0.024 L

   $$0.024 \text{ L solution} \times \frac{1.0 \text{ mol HCl}}{1.00 \text{ L solution}} = 0.024 \text{ mol HCl}$$

   Molar mass of HCl = 36.46 g

   $$0.024 \text{ mol HCl} \times \frac{36.46 \text{ g}}{1 \text{ mol}} = 0.88 \text{ g HCl}$$

48. Molar mass of NaCl = 58.44 g

   $$5.2 \text{ g NaCl} \times \frac{1 \text{ mol}}{58.44 \text{ g}} = 0.089 \text{ mol NaCl}$$

   $$0.089 \text{ mol NaCl} \times \frac{1.00 \text{ L solution}}{0.25 \text{ mol NaCl}} = 0.36 \text{ L solution can be prepared.}$$

50. a. 10.2 mL = 0.0102 L

   $$0.0102 \text{ L} \times \frac{0.451 \text{ mol AlCl}_3}{1.00 \text{ L}} \times \frac{1 \text{ mol Al}^{3+}}{1 \text{ mol AlCl}_3} = 4.60 \times 10^{-3} \text{ mol Al}^{3+}$$

188   Chapter 15   Solutions

$$0.0102 \text{ L} \times \frac{0.451 \text{ mol AlCl}_3}{1.00 \text{ L}} \times \frac{3 \text{ mol Cl}^-}{1 \text{ mol AlCl}_3} = 1.38 \times 10^{-2} \text{ mol Cl}^-$$

b. $$5.51 \text{ L} \times \frac{0.103 \text{ mol Na}_3\text{PO}_4}{1.00 \text{ L}} \times \frac{3 \text{ mol Na}^+}{1 \text{ mol Na}_3\text{PO}_4} = 1.70 \text{ mol Na}^+$$

$$5.51 \text{ L} \times \frac{0.103 \text{ mol Na}_3\text{PO}_4}{1.00 \text{ L}} \times \frac{1 \text{ mol PO}_4^{3-}}{1 \text{ mol Na}_3\text{PO}_4} = 0.568 \text{ mol PO}_4^{3-}$$

c. 1.75 mL = 0.00175 L

$$0.00175 \text{ L} \times \frac{1.25 \text{ mol CuCl}_2}{1.00 \text{ L}} \times \frac{1 \text{ mol Cu}^{2+}}{1 \text{ mol CuCl}_2} = 2.19 \times 10^{-3} \text{ mol Cu}^{2+}$$

$$0.00175 \text{ L} \times \frac{1.25 \text{ mol CuCl}_2}{1.00 \text{ L}} \times \frac{2 \text{ mol Cl}^-}{1 \text{ mol CuCl}_2} = 4.38 \times 10^{-3} \text{ mol Cl}^-$$

d. 25.2 mL = 0.0252 L

$$0.0252 \text{ L} \times \frac{0.00157 \text{ mol Ca(OH)}_2}{1.00 \text{ L}} \times \frac{1 \text{ mol Ca}^{2+}}{1 \text{ mol Ca(OH)}_2} = 3.96 \times 10^{-5} \text{ mol Ca}^{2+}$$

$$0.0252 \text{ L} \times \frac{0.00157 \text{ mol Ca(OH)}_2}{1.00 \text{ L}} \times \frac{2 \text{ mol OH}^-}{1 \text{ mol Ca(OH)}_2} = 7.91 \times 10^{-5} \text{ mol OH}^-$$

52. Molar mass of Na_2H_2EDTA = 336.2 g   250 mL = 0.250 L

$$0.250 \text{ L} \times \frac{0.0200 \text{ mol Na}_2\text{H}_2\text{EDTA}}{1.00 \text{ L}} \times \frac{336.2 \text{ g}}{1 \text{ mol}} = 1.68 \text{ g Na}_2\text{H}_2\text{EDTA}$$

54. half

56. $M_1 \times V_1 = M_2 \times V_2$

a. $M_1 = 0.251\ M$ $M_2 = ?$
    $V_1 = 125$ mL $V_2 = 250. + 125 = 375$ mL

$$M_2 = \frac{(0.251\ M)(125 \text{ mL})}{(375 \text{ mL})} = 0.0837\ M$$

b. $M_1 = 0.499\ M$ $M_2 = ?$
    $V_1 = 445$ mL $V_2 = 445 + 250. = 695$ mL

$$M_2 = \frac{(0.499\ M)(445 \text{ mL})}{(695 \text{ mL})} = 0.320\ M$$

c. $M_1 = 0.101\ M$ $M_2 = ?$
    $V_1 = 5.25$ L $V_2 = 5.25 + 0.25 = 5.50$ L

$$M_2 = \frac{(0.101\ M)(5.25 \text{ L})}{(5.50 \text{ L})} = 0.0964\ M$$

d. $M_1 = 14.5\ M$ $\qquad\qquad M_2 = ?$

$V_1 = 11.2$ mL $\qquad\qquad V_2 = 11.2 + 250. = 261.2$ mL

$$M_2 = \frac{(14.5\ M)(11.2\ \text{mL})}{(261.2\ \text{mL})} = 0.622\ \text{mL}$$

58. $M_1 = 18.1\ M$ $\qquad\qquad M_2 = 0.100\ M$

$V_1 = ?$ mL $\qquad\qquad V_2 = 125$ mL

$$M_1 = \frac{(0.100\ M)(125\ \text{mL})}{(18.1\ M)} = 0.691\ \text{mL}$$

60. $M_1 = 0.211\ M$ $\qquad\qquad M_2 = ?$

$V_1 = 75$ mL $\qquad\qquad V_2 = 125$ mL

$$M_1 = \frac{(0.211\ M)(75\ \text{mL})}{(125\ \text{mL})} = 0.127\ M = 0.13\ M$$

62. $M_1 = 6.0\ M$ $\qquad\qquad M_2 = ?$

$V_1 = 1.0$ L $\qquad\qquad V_2 = 5.0$ L $+ 1.0$ L $= 6.0$ L

$$M_1 = \frac{(6.0\ M)(1.0\ L)}{(6.0\ L)} = 1.0\ M$$

64. $Ba(NO_3)_2(aq) + Na_2SO_4(aq) \rightarrow BaSO_4(s) + 2NaNO_3(aq)$

12.5 mL = 0.0125 L

$$\text{moles Ba(NO}_3)_2 = 0.0125\ L \times \frac{0.15\ \text{mol Ba(NO}_3)_2}{1.00\ L} = 1.88 \times 10^{-3}\ \text{mol Ba(NO}_3)_2$$

From the balanced chemical equation for the reaction, if 1.88×10^{-3} mol $Ba(NO_3)_2$ are to be precipitated, then 1.88×10^{-3} mol Na_2SO_4 will be needed.

$$1.88 \times 10^{-3}\ \text{mol Na}_2SO_4 \times \frac{1.00\ L}{0.25\ \text{mol Na}_2SO_4} = 0.0075\ \text{L required} = 7.5\ \text{mL}$$

66. Molar mass $AgNO_3 = 169.91$ g $\qquad\qquad$ 135 mL = 0.135 L

$$\text{mol Cl}^- = 0.135\ L \times \frac{0.101\ \text{mol NaCl}}{1.00\ L} \times \frac{1\ \text{mol Cl}^-}{1\ \text{mol NaCl}} = 1.36 \times 10^{-2}\ \text{mol Cl}^-$$

From the balanced chemical equation for the reaction, if 1.36×10^{-2} mol of Cl^- ion is to be precipitated, then 1.36×10^{-3} mol Ag^+ is needed. Since $AgNO_3$ contains one Ag^+ per formula unit, 1.36×10^{-2} mol $AgNO_3$ is needed.

$$1.36 \times 10^{-2}\ \text{mol AgNO}_3 \times \frac{169.91\ g}{1\ \text{mol}} = 2.31\ g\ AgNO_3\ \text{is required}$$

68. 10.0 mL = 0.0100 L

$$0.0100 \text{ L} \times \frac{0.250 \text{ mol AlCl}_3}{1.00 \text{ L}} = 2.50 \times 10^{-3} \text{ mol AlCl}_3$$

AlCl$_3$(aq) + 3NaOH(s) → Al(OH)$_3$(s) + 3NaCl(aq)

$$2.50 \times 10^{-3} \text{ mol AlCl}_3 \times \frac{3 \text{ mol NaOH}}{1 \text{ mol AlCl}_3} = 7.50 \times 10^{-3} \text{ mol NaOH}$$

molar mass NaOH = 40.0 g

$$7.50 \times 10^{-3} \text{ mol NaOH} \times \frac{40.0 \text{ g NaOH}}{1 \text{ mol}} = 0.300 \text{ g NaOH}$$

70. HCl(aq) + NaOH(aq) → NaCl(aq) + H$_2$O(l)

24.9 mL = 0.0249 L

$$0.0249 \text{ L} \times \frac{0.451 \text{ mol NaOH}}{1.00 \text{ L}} = 0.0112 \text{ mol NaOH}$$

$$0.0112 \text{ mol NaOH} \times \frac{1 \text{ mol HCl}}{1 \text{ mol NaOH}} = 0.0112 \text{ mol HCl}$$

$$0.0112 \text{ mol HCl} \times \frac{1.00 \text{ L}}{0.175 \text{ mol HCl}} = 0.0642 \text{ L required} = 64.2 \text{ mL}$$

72. 7.2 mL = 0.0072 L

$$0.0072 \text{ L} \times \frac{2.5 \times 10^{-3} \text{ mol NaOH}}{1.00 \text{ L}} = 1.8 \times 10^{-5} \text{ mol NaOH}$$

H$^+$(aq) + OH$^-$(aq) → H$_2$O(l)

$$1.8 \times 10^{-5} \text{ mol OH}^- \times \frac{1 \text{ mol H}^+}{1 \text{ mol OH}^-} = 1.8 \times 10^{-5} \text{ mol H}^+$$

100 mL = 0.100 L

$$M = \frac{1.8 \times 10^{-5} \text{ mol H}^+}{0.100 \text{ L}} = 1.8 \times 10^{-4} \; M \; \text{H}^+(aq)$$

74. Experimentally, neutralization reactions are usually performed with volumetric glassware that is calibrated in milliliters rather than liters. For convenience in calculations for such reactions, the arithmetic is often performed in terms of *milli*liters and *milli*moles, rather than in liters and moles. 1 mmol = 0.001 mol. Note that the number of moles of solute per liter of solution, the molarity, is numerically equivalent to the number of *milli*moles of solute per *milli*liter of solution.

a. HNO$_3$(aq) + NaOH(aq) → NaNO$_3$(aq) + H$_2$O(l)

$$12.7 \text{ mL} \times \frac{0.501 \text{ mmol}}{1 \text{ mL}} = 6.36 \text{ mmol NaOH present in the sample}$$

Chapter 15   Solutions   191

$$6.36 \text{ mmol NaOH} \times \frac{1 \text{ mmol HNO}_3}{1 \text{ mmol NaOH}} = 6.36 \text{ mmol HNO}_3 \text{ required to react}$$

$$6.36 \text{ mmol HNO}_3 \times \frac{1.00 \text{ mL}}{0.101 \text{ mmol HNO}_3} = 63.0 \text{ mL HNO}_3 \text{ required}$$

b. $2HNO_3(aq) + Ba(OH)_2 \rightarrow Ba(NO_3)_2 + 2H_2O(l)$

$$24.9 \text{ mL} \times \frac{0.00491 \text{ mmol}}{1.00 \text{ mL}} = 0.122 \text{ mmol NaOH present in the sample}$$

$$0.122 \text{ mmol Ba(OH)}_2 \times \frac{2 \text{ mmol HNO}_3}{1 \text{ mmol Ba(OH)}_2} = 0.244 \text{ mmol HNO}_3 \text{ required}$$

$$0.244 \text{ mmol HNO}_3 \times \frac{1.00 \text{ mL}}{0.101 \text{ mmol HNO}_3} = 2.42 \text{ mL HNO}_3 \text{ is required}$$

c. $HNO_3(aq) + NH_3(aq) \rightarrow NH_4NO_3(aq)$

$$49.1 \text{ mL} \times \frac{0.103 \text{ mmol}}{1.00 \text{ mL}} = 5.06 \text{ mmol NH}_3 \text{ present in the sample}$$

$$5.06 \text{ mmol NH}_3 \times \frac{1 \text{ mmol HNO}_3}{1 \text{ mmol NH}_3} = 5.06 \text{ mmol HNO}_3 \text{ required}$$

$$5.06 \text{ mmol HNO}_3 \times \frac{1.00 \text{ mL}}{0.101 \text{ mmol HNO}_3} = 50.1 \text{ mL HNO}_3 \text{ required}$$

d. $KOH(aq) + HNO_3(aq) \rightarrow KNO_3(aq) + H_2O(l)$

$$1.21 \text{ L} \times \frac{0.102 \text{ mol}}{1.00 \text{ L}} = 0.123 \text{ mol KOH present in the sample}$$

$$0.123 \text{ mol KOH} \times \frac{1 \text{ mol HNO}_3}{1 \text{ mol KOH}} = 0.123 \text{ mol HNO}_3 \text{ required}$$

$$0.123 \text{ mol HNO}_3 \times \frac{1.00 \text{ L}}{0.101 \text{ mol HNO}_3} = 1.22 \text{ L HNO}_3 \text{ required}$$

76. 1 normal

78. 1.53 equivalents OH^- ion are needed to react with 1.53 equivalents of H^+ ion. By *definition*, one equivalent of OH^- ion exactly neutralizes one equivalent of H^+ ion.

80. $N = \dfrac{\text{number of equivalents of solute}}{\text{number of liters of solution}}$

a. equivalent weight NaOH = molar mass NaOH = 40.00 g

$$0.113 \text{ g NaOH} \times \frac{1 \text{ equiv NaOH}}{40.00 \text{ g}} = 2.83 \times 10^{-3} \text{ equiv NaOH}$$

10.2 mL = 0.0102 L

192   Chapter 15   Solutions

$$N = \frac{2.83 \times 10^{-3} \text{ equiv}}{0.0102 \text{ L}} = 0.277 \, N$$

b. equivalent weight $Ca(OH)_2 = \dfrac{\text{molar mass}}{2} = \dfrac{74.10 \text{ g}}{2} = 37.05 \text{ g}$

$$12.5 \text{ mg} \times \frac{1 \text{ g}}{10^3 \text{ mg}} \times \frac{1 \text{ equiv}}{37.05 \text{ g}} = 3.37 \times 10^{-4} \text{ equiv } Ca(OH)_2$$

100. mL = 0.100 L

$$N = \frac{3.37 \times 10^{-3} \text{ equiv}}{0.100 \text{ L}} = 3.37 \times 10^{-3} \, N$$

c. equivalent weight $H_2SO_4 = \dfrac{\text{molar mass}}{2} = \dfrac{98.09 \text{ g}}{2} = 49.05 \text{ g}$

$$12.4 \text{ g} \times \frac{1 \text{ equiv}}{49.05 \text{ g}} = 0.253 \text{ equiv } H_2SO_4$$

155 mL = 0.155 L

$$N = \frac{0.253 \text{ equiv}}{0.155 \text{ L}} = 1.63 \, N$$

82. a. $4.01 \times 10^{-3} \, M \, H_3PO_4 \times \dfrac{3 \text{ equiv } H_3PO_4}{1 \text{ mol } H_3PO_4} = 1.20 \times 10^{-2} \, N \, H_3PO_4$

b. $0.000101 \, M \, Ba(OH)_2 \times \dfrac{2 \text{ equiv } Ba(OH)_2}{1 \text{ mol } Ba(OH)_2} = 2.02 \times 10^{-4} \, N \, Ba(OH)_2$

c. $0.101 \, M \, HC_2H_3O_2 \times \dfrac{1 \text{ equiv } HC_2H_3O_2}{1 \text{ mol } HC_2H_3O_2} = 0.101 \, N \, HC_2H_3O_2$

84. Molar mass of $Ca(OH)_2$ = 74.10 g

$$5.21 \text{ mg } Ca(OH)_2 \times \frac{1 \text{ g}}{10^3 \text{ mg}} \times \frac{1 \text{ mol}}{74.10 \text{ g}} = 7.03 \times 10^{-5} \text{ mol } Ca(OH)_2$$

1000. mL = 1.000 L (volumetric flask volume: 4 significant figures).

$$M = \frac{7.03 \times 10^{-5} \text{ mol}}{1.000 \text{ L}} = 7.03 \times 10^{-5} \, M \, Ca(OH)_2$$

$$N = 7.03 \times 10^{-5} \, M \, Ca(OH)_2 \times \frac{2 \text{ equiv } Ca(OH)_2}{1 \text{ mol } Ca(OH)_2} = 1.41 \times 10^{-4} \, N \, Ca(OH)_2$$

86. $H_2SO_4(aq) + 2NaOH \rightarrow Na_2SO_4(aq) + 2H_2O(l)$

0.103 $M \, H_2SO_4$ = 0.206 $N \, H_2SO_4$

0.104 M NaOH = 0.104 N NaOH

$$4.25 \text{ L NaOH} \times \frac{0.104 \text{ equiv}}{1.00 \text{ L}} = 0.442 \text{ equiv NaOH}$$

0.442 equiv NaOH requires, by definition, 0.442 equiv H_2SO_4 to react.

$$0.442 \text{ equiv } H_2SO_4 \times \frac{1.00 \text{ L}}{0.206 \text{ equiv}} = 2.15 \text{ L } H_2SO_4 = 2.15 \times 10^3 \text{ mL } H_2SO_4$$

88. $2NaOH(aq) + H_2SO_4(aq) \rightarrow Na_2SO_4(aq) + 2 H_2O(l)$

$$27.34 \text{ mL NaOH} \times \frac{0.1021 \text{ mmol}}{1.00 \text{ mL}} = 2.791 \text{ mmol NaOH}$$

$$2.791 \text{ mmol NaOH} \times \frac{1 \text{ mmol } H_2SO_4}{2 \text{ mmol NaOH}} = 1.396 \text{ mmol } H_2SO_4$$

$$M = \frac{1.396 \text{ mmol } H_2SO_4}{25.00 \text{ mL}} = 0.05583 \text{ } M \text{ } H_2SO_4 = 0.1117 \text{ } N \text{ } H_2SO_4$$

90. Molarity is defined as the number of moles of solute contained in 1 liter of *total* solution volume (solute plus solvent after mixing). In the first case, where 50. g of NaCl is dissolved in 1.0 L of water, the total volume after mixing is *not* known and the molarity cannot be calculated. In the second example, the final volume after mixing is known and the molarity can be calculated simply.

92. $$75 \text{ g solution} \times \frac{25 \text{ g NaCl}}{100 \text{ g solution}} = 18.75 \text{ g NaCl}$$

$$18.75 \text{ g NaCl} \times \frac{100 \text{ g solution}}{5.0 \text{ g NaCl}} = 375 \text{ g} = 3.8 \times 10^2 \text{ g}$$

94. $Ba(NO_3)_2(aq) + H_2SO_4(aq) \rightarrow BaSO_4(s) + 2HNO_3(aq)$

    10. mL = 0.010 L

$$0.010 \text{ L Ba(NO}_3)_2 \times \frac{0.50 \text{ mol Ba(NO}_3)_2}{1.00 \text{ L}} = 0.0050 \text{ mol Ba(NO}_3)_2$$

$$0.010 \text{ L } H_2SO_4 \times \frac{0.20 \text{ mol } H_2SO_4}{1.00 \text{ L}} = 0.0020 \text{ mol } H_2SO_4$$

Since the coefficients of $Ba(NO_3)_2$ and H_2SO_4 in the balanced chemical equation for the reaction are both *one*, then the 0.0020 mol H_2SO_4 is the limiting reactant for the precipitation reaction. Only 0.0020 mol of Ba^{2+} ion will precipitate as $BaSO_4(s)$, leaving (0.0050 − 0.0020) = 0.0030 mol of Ba^{2+} ion remaining in solution. This quantity of Ba^{2+} ion left in solution is now dispersed through the *combined* volume of solution (20 mL = 0.020 L).

molar mass $BaSO_4$ = 233.4 g

194   Chapter 15   Solutions

$$0.0020 \text{ mol BaSO}_4 \times \frac{233.4 \text{ g BaSO}_4}{1 \text{ mol BaSO}_4} = 0.47 \text{ g BaSO}_4 \text{ precipitate}$$

$$M_{\text{barium remaining}} = \frac{0.0030 \text{ mol Ba}^{2+}}{0.020 \text{ L}} = 0.15 \text{ M Ba}^{2+} \text{ in solution}$$

96. molar mass H_2O = 18.0 g

1.0 L water = 1.0×10^3 mL water ≈ 1.0×10^3 g water

$$1.0 \times 10^3 \text{ g H}_2\text{O} \times \frac{1 \text{ mol H}_2\text{O}}{18.0 \text{ g H}_2\text{O}} = 56 \text{ mol H}_2\text{O}$$

98. 500 mL HCl solution = 0.500 L HCl solution

$$0.500 \text{ L solution} \times \frac{0.100 \text{ mol HCl}}{1.00 \text{ L HCl solution}} = 0.0500 \text{ mol HCl}$$

$$0.0500 \text{ mol HCl} \times \frac{22.4 \text{ L HCl gas at STP}}{1 \text{ mol HCl}} = 1.12 \text{ L HCl gas at STP}$$

100. $$10.0 \text{ g HCl} \times \frac{100. \text{ g solution}}{33.1 \text{ g HCl}} = 30.21 \text{ g solution}$$

$$30.21 \text{ g solution} \times \frac{1.00 \text{ mL solution}}{1.147 \text{ g solution}} = 26.3 \text{ mL solution}$$

102. molar mass NaCl = 58.44 g

$$10.0 \text{ g NaCl} \times \frac{1 \text{ mol NaCl}}{58.44 \text{ g NaCl}} = 0.1711 \text{ mol NaCl}$$

100. mL = 0.100 L

$$M = \frac{0.1711 \text{ mol NaCl}}{0.100 \text{ L}} = 1.71 \text{ M}$$

104. a. $$\frac{5.0 \text{ g KNO}_3}{(5.0 \text{ g KNO}_3 + 75 \text{ g H}_2\text{O})} \times 100 = \frac{5.0 \text{ g}}{80.0 \text{ g}} \times 100 = 6.3\% \text{ KNO}_3$$

b. 2.5 mg = 0.0025 g

$$\frac{0.0025 \text{ g KNO}_3}{(0.025 \text{ g KNO}_3 + 1.0 \text{ g H}_2\text{O})} \times 100 = \frac{0.0025 \text{ g}}{1.0025 \text{ g}} \times 100 = 0.25\% \text{ KNO}_3$$

c. $$\frac{11 \text{ g KNO}_3}{(11 \text{ g KNO}_3 + 89 \text{ g H}_2\text{O})} \times 100 = \frac{11 \text{ g}}{100 \text{ g}} \times 100 = 11\% \text{ KNO}_3$$

d. $$\frac{11 \text{ g KNO}_3}{(11 \text{ g KNO}_3 + 49 \text{ g H}_2\text{O})} \times 100 = \frac{11 \text{ g}}{60 \text{ g}} \times 100 = 18\% \text{ KNO}_3$$

106. $$\%C = \frac{5.0 \text{ g C}}{(5.0 \text{ g C} + 1.5 \text{ g Ni} + 100. \text{ g Fe})} \times 100 = \frac{5.0 \text{ g}}{106.5 \text{ g}} \times 100 = 4.7\% \text{ C}$$

$$\%\text{Ni} = \frac{1.5 \text{ g Ni}}{(5.0 \text{ g C} + 1.5 \text{ g Ni} + 100.\text{ g Fe})} \times 100 = \frac{1.5 \text{ g}}{106.5 \text{ g}} \times 100 = 1.4\% \text{ Ni}$$

$$\%\text{Fe} = \frac{100.\text{ g Fe}}{(5.0 \text{ g C} + 1.5 \text{ g Ni} + 100.\text{ g Fe})} \times 100 = \frac{100.\text{ g}}{106.5 \text{ g}} \times 100 = 93.9\% \text{ Fe}$$

108. To say that the solution is 5.5% by mass Na_2CO_3 means that 5.5 g of Na_2CO_3 are contained in every 100 g of the solution.

$$500.\text{ g solution} \times \frac{5.5 \text{ g Na}_2\text{CO}_3}{100 \text{ g solution}} = 28 \text{ g Na}_2\text{CO}_3$$

110. For NaCl: $125 \text{ g solution} \times \dfrac{7.5 \text{ g NaCl}}{100 \text{ g solution}} = 9.4 \text{ g NaCl}$

 For KBr: $125 \text{ g solution} \times \dfrac{2.5 \text{ g KBr}}{100 \text{ g solution}} = 3.1 \text{ g KBr}$

112. Molarity $= \dfrac{\text{moles of solute}}{\text{liters of solution}}$

 a. 25 mL = 0.025 L

 $$M = \frac{0.10 \text{ mol CaCl}_2}{0.025 \text{ L solution}} = 4.0 \text{ }M$$

 b. $M = \dfrac{2.5 \text{ mol KBr}}{2.5 \text{ L solution}} = 1.0 \text{ }M$

 c. 755 mL = 0.755 L

 $$M = \frac{0.55 \text{ mol NaNO}_3}{0.755 \text{ L solution}} = 0.73 \text{ }M$$

 d. $M = \dfrac{4.5 \text{ mol Na}_2\text{SO}_4}{1.25 \text{ L solution}} = 3.6 \text{ }M$

114. molar mass $C_{12}H_{22}O_{11}$ = 342.3 g

 $$125 \text{ g } C_{12}H_{22}O_{11} \times \frac{1 \text{ mol}}{342.3 \text{ g}} = 0.3652 \text{ mol } C_{12}H_{22}O_{11}$$

 450. mL = 0.450 L

 $$M = \frac{0.3652 \text{ mol } C_{12}H_{22}O_{11}}{0.450 \text{ L solution}} = 0.812 \text{ }M$$

116. molar mass NaCl = 58.44 g

 $$1.5 \text{ g NaCl} \times \frac{1 \text{ mol}}{58.44 \text{ g}} = 0.0257 \text{ mol NaCl}$$

196   Chapter 15   Solutions

$$M = \frac{0.0257 \text{ mol NaCl}}{1.0 \text{ L solution}} = 0.026 \ M$$

118. a. $3.8 \text{ L solution} \times \dfrac{1.5 \text{ mol KCl}}{1.00 \text{ L solution}} = 5.7 \text{ mol KCl}$

molar mass KCl = 74.6 g

$5.7 \text{ mol KCl} \times \dfrac{74.6 \text{ g KCl}}{1 \text{ mol KCl}} = 420 \text{ g KCl} = 4.2 \times 10^2 \text{ g KCl}$

b.   15 mL = 0.015 L

$0.015 \text{ L solution} \times \dfrac{5.4 \text{ mol NaCl}}{1.00 \text{ L solution}} = 0.0810 \text{ mol NaCl}$

molar mass NaCl = 58.44 g

$0.0810 \text{ mol NaCl} \times \dfrac{58.44 \text{ g NaCl}}{1 \text{ mol NaCl}} = 4.7 \text{ g NaCl}$

c. $20. \text{ L solution} \times \dfrac{12.1 \text{ mol HCl}}{1.00 \text{ L solution}} = 242 \text{ mol HCl}$

molar mass HCl = 36.46 g

$242 \text{ mol HCl} \times \dfrac{36.46 \text{ g HCl}}{1 \text{ mol HCl}} = 8823 \text{ g HCl} = 8.8 \times 10^3 \text{ g HCl}$

d.   25 mL = 0.025 L

$0.025 \text{ L solution} \times \dfrac{0.100 \text{ mol HClO}_4}{1.00 \text{ L solution}} = 0.00250 \text{ mol HClO}_4$

molar mass $HClO_4$ = 100.5 g

$0.00250 \text{ mol HClO}_4 \times \dfrac{100.5 \text{ g HClO}_4}{1 \text{ mol HClO}_4} = 0.25 \text{ g HClO}_4$

120. a. $1.25 \text{ L} \times \dfrac{0.250 \text{ mol Na}_3\text{PO}_4}{1.00 \text{ L}} = 0.3125 \text{ mol Na}_3\text{PO}_4 =$

$0.3125 \text{ mol Na}_3\text{PO}_4 \times \dfrac{3 \text{ mol Na}^+}{1 \text{ mol Na}_3\text{PO}_4} = 0.938 \text{ mol Na}^+$

$0.3125 \text{ mol Na}_3\text{PO}_4 \times \dfrac{1 \text{ mol PO}_4^{3-}}{1 \text{ mol Na}_3\text{PO}_4} = 0.313 \text{ mol PO}_4^{3-}$

b.   3.5 mL = 0.0035 L

$0.0035 \text{ L} \times \dfrac{6.0 \text{ mol H}_2\text{SO}_4}{1.00 \text{ L}} = 0.021 \text{ mol H}_2\text{SO}_4$

$$0.021 \text{ mol H}_2\text{SO}_4 \times \frac{2 \text{ mol H}^+}{1 \text{ mol H}_2\text{SO}_4} = 0.042 \text{ mol H}^+$$

$$0.021 \text{ mol H}_2\text{SO}_4 \times \frac{1 \text{ mol SO}_4^{2-}}{1 \text{ mol H}_2\text{SO}_4} = 0.021 \text{ mol SO}_4^{2-}$$

c. 25 mL = 0.025 L

$$0.025 \text{ L} \times \frac{0.15 \text{ mol AlCl}_3}{1.00 \text{ L}} = 0.00375 \text{ mol AlCl}_3$$

$$0.00375 \text{ mol AlCl}_3 \times \frac{1 \text{ mol Al}^{3+}}{1 \text{ mol AlCl}_3} = 0.0038 \text{ mol Al}^{3+}$$

$$0.00375 \text{ mol AlCl}_3 \times \frac{3 \text{ mol Cl}^-}{1 \text{ mol AlCl}_3} = 0.011 \text{ mol Cl}^-$$

d. $$1.50 \text{ L} \times \frac{1.25 \text{ mol BaCl}_2}{1.00 \text{ L}} = 1.875 \text{ mol BaCl}_2$$

$$1.875 \text{ mol BaCl}_2 \times \frac{1 \text{ mol Ba}^{2+}}{1 \text{ mol BaCl}_2} = 1.88 \text{ mol Ba}^{2+}$$

$$1.875 \text{ mol BaCl}_2 \times \frac{2 \text{ mol Cl}^-}{1 \text{ mol BaCl}_2} = 3.75 \text{ mol Cl}^-$$

122. $M_1 \times V_1 = M_2 \times V_2$

a. $M_1 = 0.200\ M$ $M_2 = ?$
    $V_1 = 125$ mL $V_2 = 125 + 150. = 275$ mL

$$M_2 = \frac{(0.200\ M)(125 \text{ mL})}{(275 \text{ mL})} = 0.0909\ M$$

b. $M_1 = 0.250\ M$ $M_2 = ?$
    $V_1 = 155$ mL $V_2 = 155 + 150. = 305$ mL

$$M_2 = \frac{(0.250\ M)(155 \text{ mL})}{(305 \text{ mL})} = 0.127\ M$$

c. $M_1 = 0.250\ M$ $M_2 = ?$
    $V_1 = 0.500$ L = 500. mL $V_2 = 500. + 150. = 650.$ mL

$$M_2 = \frac{(0.250\ M)(500. \text{ mL})}{(650 \text{ mL})} = 0.192\ M$$

d. $M_1 = 18.0\ M$ $M_2 = ?$
    $V_1 = 15$ mL $V_2 = 15 + 150. = 165$ mL

$$M_2 = \frac{(18.0\ M)(15 \text{ mL})}{(165 \text{ mL})} = 1.6\ M$$

198 Chapter 15   Solutions

124. $M_1 \times V_1 = M_2 \times V_2$

$M_1 = 5.4\ M$ $\quad\quad\quad\quad\quad\quad\quad\quad\quad M_2 = ?$

$V_1 = 50.\ mL$ $\quad\quad\quad\quad\quad\quad\quad\quad V_2 = 300.\ mL$

$$M_2 = \frac{(5.4\ M)(50.\ mL)}{(300.\ mL)} = 0.90\ M$$

126. $25.0\ mL = 0.0250\ L$

$$0.0250\ L\ NiCl_2\ solution \times \frac{0.20\ mol\ NiCl_2}{1.00\ L\ NiCl_2\ solution} = 0.00500\ mol\ NiCl_2$$

$$0.00500\ mol\ NiCl_2 \times \frac{1\ mol\ Na_2S}{1\ mol\ NiCl_2} = 0.00500\ mol\ Na_2S$$

$$0.00500\ mol\ Na_2S \times \frac{1.00\ L\ Na_2S\ solution}{0.10\ mol\ Na_2S} = 0.050\ L = 50.\ mL\ Na_2S\ solution$$

128. $HNO_3(aq) + NaOH(aq) \rightarrow NaNO_3(aq) + H_2O(l)$

$35.0\ mL = 0.0350\ L$

$$0.0350\ L \times \frac{0.150\ mol\ NaOH}{1.00\ L} = 5.25 \times 10^{-3}\ mol\ NaOH$$

$$5.25 \times 10^{-3}\ mol\ NaOH \times \frac{1\ mol\ HNO_3}{1\ mol\ NaOH} = 5.25 \times 10^{-3}\ mol\ HNO_3$$

$$5.25 \times 10^{-3}\ mol\ HNO_3 \times \frac{1.00\ L}{0.150\ mol\ HNO_3} = 0.0350\ L = 35.0\ mL\ HNO_3$$

130. $N = \dfrac{\text{number of equivalents of solute}}{\text{number of liters of solution}}$

a. equivalent weight HCl = molar mass HCl = 36.46 g

$$15.0\ g\ HCl \times \frac{1\ equiv\ HCl}{36.46\ g\ HCl} = 0.411\ equiv\ HCl$$

$500.\ mL = 0.500\ L$

$$N = \frac{0.411\ equiv}{0.500\ L} = 0.822\ N$$

b. equivalent weight $H_2SO_4 = \dfrac{molar\ mass}{2} = \dfrac{98.09\ g}{2} = 49.05\ g$

$$49.0\ g\ H_2SO_4 \times \frac{1\ equiv\ H_2SO_4}{49.05\ g\ H_2SO_4} = 0.999\ equiv\ H_2SO_4$$

$250.\ mL = 0.250\ L$

$$N = \frac{0.999 \text{ equiv}}{0.250 \text{ L}} = 4.00 \ N$$

c. equivalent weight H_3PO_4 = $\dfrac{\text{molar mass}}{3}$ = $\dfrac{98.0 \text{ g}}{3}$ = 32.67 g

$$10.0 \text{ g } H_3PO_4 \times \frac{1 \text{ equiv } H_3PO_4}{32.67 \text{ g } H_3PO_4} = 0.3061 \text{ equiv } H_3PO_4$$

100. mL = 0.100 L

$$N = \frac{0.3061 \text{ equiv}}{0.100 \text{ L}} = 3.06 \ N$$

132. molar mass NaH_2PO_4 = 120.0 g

$$5.0 \text{ g } NaH_2PO_4 \times \frac{1 \text{ mol } NaH_2PO_4}{120.0 \text{ g } NaH_2PO_4} = 0.04166 \text{ mol } NaH_2PO_4$$

500. mL = 0.500 L

$$M = \frac{0.04166 \text{ mol}}{0.500 \text{ L}} = 0.08333 \ M \ NaH_2PO_4 = 0.083 \ M \ NaH_2PO_4$$

$$0.08333 \ M \ NaH_2PO_4 \times \frac{2 \text{ equiv } NaH_2PO_4}{1 \text{ mol } NaH_2PO_4} = 0.1667 \ N \ NaH_2PO_4 = 0.17 \ N \ NaH_2PO_4$$

134. $N_{acid} \times V_{acid} = N_{base} \times V_{base}$
$N_{acid} \times (10.0 \text{ mL}) = (3.5 \times 10^{-2} \ N)(27.5 \text{ mL})$
$N_{acid} = 9.6 \times 10^{-2} \ N \ HNO_3$

Cumulative Review: Chapters 13, 14, and 15

2. The pressure exerted by the atmosphere is due to the several mile thick layer of gases above the surface of the earth pressing down on us. Atmospheric pressure has traditionally been measured with a mercury barometer (see Figure 13.2). A mercury barometer usually consists of a glass tube which is sealed at one end and filled with mercury. The tube is then inverted over an open reservoir also containing mercury. When the tube is inverted, most of the mercury does not fall out of the tube. Since the reservoir of mercury is open to the atmosphere, the atmospheric pressure on the surface of the mercury in the reservoir is enough pressure to hold the bulk of the mercury in the glass tube. The pressure of the atmosphere is sufficient, on average, to support a column of mercury 76 cm (760 mm) high in the tube.

4. Boyle's law basically says that the volume of a gas sample will decrease if you squeeze harder on it. Imagine squeezing hard on a tennis ball with your hand: the ball collapses as the gas inside it is forced into a smaller volume by your hand. Of course, to be perfectly correct, the temperature and amount of gas (moles) must remain the same while you adjust the pressure for Boyle's law to hold true. The first of the two mathematical statements of Boyle's law you should remember is

$$P \times V = \text{constant}$$

which basically is the definition of Boyle's law (in order for the product $(P \times V)$ to remain constant, if one of these terms increases the other must decrease). The second formulation of Boyle's law you have to be able to deal with is the one more commonly used in solving problems,

$$P_1 \times V_1 = P_2 \times V_2$$

With this second formulation, we can determine pressure-volume information about a given sample under two sets of conditions. These two mathematical formulas are just two different ways of saying the same thing: if the pressure on a sample of gas is increased, the volume of the sample of gas will decrease. A graph of Boyle's law data is given as Figure 13.5: this sort of graph ($xy = k$) is known to mathematicians as a hyperbola.

6. Charles's law basically says that if you heat a sample of gas, the volume of the sample will increase. That is, when the temperature of a gas is increased, the volume of the gas also increases (assuming the pressure and amount of gas remains the same). Charles's law is a direct proportionality (if you increase T, this increases V), whereas Boyle's law is an inverse proportionality (if you increase P, this decreases V). There are two mathematical statements of Charles's law you should be familiar with. The first statement

$$V = bT$$

is just a definition (the volume of a gas sample is directly related to its temperature: if you increase the temperature, the volume increases). The working formulation of Charles's law we use in problem solving is given as

$$\frac{V_1}{T_1} = \frac{V_2}{T_2}$$

With this formulation, we can determine volume-temperature information for a given gas sample under two sets of conditions. Charles's law only holds true if the amount of gas remains the same (obviously the volume of a gas sample would increase if there were more gas present) and also if the pressure remains the same (a change in pressure also changes the volume of a gas sample).

8. Avogadro's law tells us that, with all other things being equal, two moles of gas is twice as big as one mole of gas! That is, the volume of a sample of gas is directly proportional to the amount of gas present (at constant temperature and pressure). If we want to compare the volumes of two samples of the same gas as an indication of the amount of gas present in the samples, we would have to make certain that the two samples of gas are at the same pressure and temperature: the volume of a sample of gas would vary with either temperature or pressure, or both. Avogadro's law holds true for comparing gas samples that are under the same conditions. Avogadro's law is a direct proportionality: the more gas you have in a sample, the larger the sample's volume will be.

10. The "partial" pressure of an individual gas in a mixture of gases represents the pressure the gas would have in the same container at the same temperature if it were the only gas present. The total pressure in a mixture of gases is just the sum of the individual partial pressures of the gases present in the mixture. Because the partial pressures of the gases in a mixture are additive (i.e., the total pressure is the sum of the partial pressures), this suggests that the total pressure in a container is a function really only of the number of molecules present in the same, and not of the identity of the molecules or any other property of the molecules (such as their inherent atomic size).

12. The main postulates of the kinetic-molecular theory for gases are as follows: (a) gases consist of tiny particles (atoms or molecules), and the size of these particles themselves is negligible compared to the bulk volume of a gas sample; (b) the particles in a gas are in constant random motion, colliding with each other and with the walls of the container; (c) the particles in a gas sample do not assert any attractive or repulsive forces on one another; (d) the average kinetic energy of the particles in a sample of gas is directly related to the absolute temperature of the gas sample. The pressure exerted by a gas is a result of the molecules colliding with (and pushing on) the walls of the container; the pressure increases with temperature because at a higher temperature, the molecules are moving faster and hit the walls of the container with greater force. A gas fills whatever volume is available to it because the molecules in a gas are in constant random motion: if the motion of the molecules is random, they eventually will move out into whatever volume is available until the distribution of molecules is uniform; at constant pressure, the volume of a gas sample increases as the temperature is increased because with each collision

having greater force, the container must expand so that the molecules (and therefore the collisions) are farther apart if the pressure is to remain constant.

14. Solids and liquids are much more condensed states of matter than are gases: the molecules are much closer together in solids and liquids and interact with each other to a much greater extent. Solids and liquids have much greater densities than do gases, and are much less compressible, because there is so little room between the molecules in the solid and liquid states (solids and liquids effectively have native volumes of their own, and their volumes are not affected nearly as much by the temperature or pressure). Although solids are more rigid than liquids, the solid and liquid state have much more in common with each other than either of these states has with the gaseous state. We know this is true since it typically only takes a few kilojoules of energy to melt 1 mol of a solid (since not much change has to take place in the molecules), whereas it may take 10 times more energy to vaporize a liquid (since there is such a great change between the liquid and gaseous states).

16. The normal boiling point of water, that is, water's boiling point at a pressure of exactly 760 mm Hg, is 100°C (you will recall that the boiling point of water was used to set one of the reference temperatures of the Celsius temperature scale). Water remains at 100°C while boiling, until all the water has boiled away, because the additional heat energy being added to the sample is used to overcome attractive forces among the water molecules as they go from the condensed, liquid state to the gaseous state. The normal (760 mm Hg) freezing point of water is exactly 0°C (again, this property of water was used as one of the reference points for the Celsius temperature scale). A cooling curve for water is given as Figure 14.2: notice how the curve shows that the amount of heat needed to boil the sample is much larger than the amount needed to melt the sample.

18. Dipole-dipole forces are a type of intermolecular force that can exist between molecules with permanent dipole moments. Molecules with permanent dipole moments try to orient themselves so that the positive end of one polar molecule can attract the negative end of another polar molecule. Dipole-dipole forces are not nearly as strong as ionic or covalent bonding forces (only about 1% strong as covalent bonding forces) since electrostatic attraction is related to the magnitude of the charges of the attracting species. Since polar molecules have only a "partial" charge at each end of the dipole, the magnitude of the attractive force is not as large. The strength of such forces also drops rapidly as molecules become farther apart and is important only in the solid and liquid states (such forces are negligible in the gaseous state since the molecules are too far apart). Hydrogen bonding is an especially strong sort of dipole-dipole attractive force which can exist when hydrogen atoms are directly bonded to the most strongly electronegative atoms (N, O, and F). Because the hydrogen atom is so small, dipoles involving N-H, O-H, and F-H bonds can approach each other much more closely than can dipoles involving other atoms. Since the

magnitude of dipole-dipole forces is dependent on distance, unusually strong attractive forces can exist in such molecules. We take the fact that the boiling point of water is so much higher than that of the other covalent hydrogen compounds of the Group 6 elements as evidence for the special strength of hydrogen bonding (it takes more energy to vaporize water because of the extra strong forces holding together the molecules in the liquid state).

20. Vaporization of a liquid requires and input of energy because the intermolecular forces which hold the molecules together in the liquid state must be overcome. The high heat of vaporization of water is essential to life on earth since much of the excess energy striking the earth from the sun is dissipated in vaporizing water. Condensation is the opposite process to vaporization: that is, condensation refers to the process by which molecules in the vapor state form a liquid. In a closed container containing a liquid and some empty space above the liquid, an equilibrium is set up between vaporization and condensation. The liquid in such a sealed container never completely evaporates: when the liquid is first placed in the container, the liquid phase begins to evaporate into the empty space. As the number of molecules in the vapor phase begins to get large, however, some of these molecules begin to re-enter the liquid phase. Eventually, every time a molecule of liquid somewhere in the container enters the vapor phase, somewhere else in the container a molecule of vapor re-enters the liquid. There is no further net change in the amount of liquid phase (although molecules are continually moving between the liquid and vapor phases). The pressure of the vapor in such an equilibrium situation is characteristic for the liquid at each particular temperature (for example, the vapor pressures of water are tabulated at different temperatures in Table 13.2). A simple experiment to determine vapor pressure is shown in Figure 14.10: samples of a liquid are injected into a sealed tube containing mercury; since mercury is so dense, the liquids float to the top of the mercury where they evaporate; as the vapor pressures of the liquids develop to the saturation point, the level of mercury in the tube changes as an index of the magnitude of the vapor pressures. Typically, liquids with strong intermolecular forces have small vapor pressures (they have more difficulty in evaporating) than do liquids with very weak intermolecular forces: for example, the components of gasoline (weak forces) have much higher vapor pressures than does water (strong forces) and evaporate more easily.

22. The simple model we use to explain many properties of metallic elements is called the electron sea model. In this model we picture a regular lattice array of metal cation in sort of a "sea" of mobile valence electrons. The electrons can move easily to conduct heat or electricity through the metal, and the lattice of cations can be deformed fairly easily, allowing the metal to be hammered into a sheet or stretched to make a wire. An alloy is a material that contains a mixture of elements, which overall has metallic properties. Substitutional alloys consist of a host metal in which some of the atoms in the metal's crystalline structure are replaced by atoms of of other metallic elements of comparable size to the atoms of the host metal. For example, sterling

24. A saturated solution is one that contains as much solute as can dissolve at a particular temperature. To say that a solution is saturated doe not necessarily mean that the solute is present at a high concentration: for example, magnesium hydroxide only dissolves to a very small extent before the solution is saturated, whereas it takes a great deal of sugar to form a saturated solution (and the saturated solution is extremely concentrated). A saturated solution is one which is in equilibrium with undissolved solute: as molecules of solute dissolve from the solid in one place in the solution, dissolved molecules rejoin the solid phase in another place in the solution. As with the development of vapor pressure above a liquid (see Question 20 above), formation of a solution reaches a state of dynamic equilibrium: once the rates of dissolving and "undissolving" become equal, there will be no further net change in the concentration of the solution and the solution will be saturated.

26. Adding additional solvent to a solution so as to dilute the solution *does not change* the number of moles of solute present, but only changes the volume in which the solute is dispersed. If we are using the molarity of the solution to describe its concentration, the number of liters is changed when we add solvent, and the number of moles per liter (the molarity) changes, but the actual number of moles of solute does not change. For example, 125 mL of 0.551 M NaCl contains 68.9 millimol of NaCl. The solution will still contain 68.9 millimol of NaCl after the 250 mL of water is added to it, only now the 68.9 millimol of NaCl will be dispersed in a total volume of 375 mL. This gives the new molarity as 68.9 mmol/375 mL = 0.184 M. The volume and the concentration have changed, but the number of moles of solute in the solution has not changed.

28. $$P_2 = \frac{P_1 \times V_1}{V_2}$$

a. 701 torr = 701 mm Hg

$$P_2 = \frac{(761 \text{ mm Hg})(24.5 \text{ mL})}{(701 \text{ mm Hg})} = 26.6 \text{ mL}$$

b. 0.549 atm = 417.2 mm Hg

$$P_2 = \frac{(761 \text{ mm Hg})(5.41 \text{ L})}{(417.2 \text{ mm Hg})} = 9.87 \text{ L}$$

c. 55.1 kPa = 413.3 mm Hg

$$P_2 = \frac{(761 \text{ mm Hg})(221 \text{ mL})}{(413.3 \text{ mm Hg})} = 407 \text{ mL}$$

30. a. $P_1 = 775$ mm Hg $\qquad P_2 = 760$ mm Hg
    $V_1 = 45.1$ mL $\qquad V_2 = ?$
    $T_1 = 24.1°C = 297.1$ K $\qquad T_2 = 273$ K

$$V_2 = \frac{P_1 V_1 T_2}{T_1 P_2} = \frac{(775 \text{ mm Hg})(45.1 \text{ mL})(273 \text{ K})}{(297.1 \text{ K})(760 \text{ mm Hg})} = 42.3 \text{ mL}$$

b. $P_1 = 0.890$ atm $\qquad P_2 = 1.00$ atm

   $V_1 = 4.31$ L $\qquad V_2 = ?$

   $T_1 = 72.1°C = 345.1$ K $\qquad T_2 = 273$ K

$$V_2 = \frac{P_1 V_1 T_2}{T_1 P_2} = \frac{(0.890 \text{ atm})(4.31 \text{ L})(273 \text{ K})}{(345.1 \text{ K})(1.00 \text{ atm})} = 3.03 \text{ L}$$

c. $P_1 = 91.2$ kPa $\qquad P_2 = 101.325$ kPa

   $V_1 = 5.12$ mL $\qquad V_2 = ?$

   $T_1 = 289$ K $\qquad T_2 = 273$ K

$$V_2 = \frac{P_1 V_1 T_2}{T_1 P_2} = \frac{(91.2 \text{ kPa})(5.12 \text{ mL})(273 \text{ K})}{(289 \text{ K})(101.325 \text{ kPa})} = 4.35 \text{ mL}$$

d. $P_1 = 1.45$ atm $\qquad P_2 = 1.00$ atm

   $V_1 = 91.3$ L $\qquad V_2 = ?$

   $T_1 = 451°C = 724$ K $\qquad T_2 = 273$ K

$$V_2 = \frac{P_1 V_1 T_2}{T_1 P_2} = \frac{(1.45 \text{ atm})(91.3 \text{ L})(273 \text{ K})}{(724 \text{ K})(1.00 \text{ atm})} = 49.9 \text{ L}$$

32. Since we have 100. g of the solution, which is 3.11% by mass H_2O_2, then the sample must contain 3.11 g of H_2O_2

molar mass H_2O_2 = 34.02 g

$$3.11 \text{ g } H_2O_2 \times \frac{1 \text{ mol } H_2O_2}{34.02 \text{ g } H_2O_2} = 0.0914 \text{ mol } H_2O_2$$

$$0.0914 \text{ mol } H_2O_2 \times \frac{1 \text{ mol } O_2}{2 \text{ mol } H_2O_2} = 0.0457 \text{ mol } O_2 \text{ produced}$$

At STP, $0.0457 \text{ mol } O_2 \times \dfrac{22.4 \text{ L } O_2}{1 \text{ mol } O_2} = 1.02 \text{ L } O_2$

$P_1 = 760 \text{ mm Hg}$ $\qquad\qquad P_2 = 771 \text{ mm Hg}$

$V_1 = 1.02 \text{ L}$ $\qquad\qquad\quad V_2 = ?$

$T_1 = 273 \text{ K}$ $\qquad\qquad\quad\; T_2 = 294 \text{ K}$

$V_2 = \dfrac{P_1 V_1 T_2}{T_1 P_2} = \dfrac{(760 \text{ mm Hg})(1.02 \text{ L})(294 \text{ K})}{(273 \text{ K})(771 \text{ mm Hg})} = 1.10 \text{ L}$

34. a. $\dfrac{5.51 \text{ g NaCl}}{(5.51 + 25.0) \text{ g solution}} \times 100 = 18.1\% \text{ NaCl}$

　　b. $\dfrac{5.51 \text{ g NaCl}}{(5.51 + 45.1) \text{ g solution}} \times 100 = 10.9\% \text{ NaCl}$

　　c. $\dfrac{5.51 \text{ g NaCl}}{(5.51 + 100.) \text{ g solution}} \times 100 = 5.22\% \text{ NaCl}$

　　d. $\dfrac{5.51 \text{ g NaCl}}{(5.51 + 995) \text{ g solution}} \times 100 = 0.551\% \text{ NaCl}$

　　e. $\dfrac{5.51 \text{ g NaCl}}{(5.51 + 5034) \text{ g solution}} \times 100 = 0.109\% \text{ NaCl}$

36. a. $\dfrac{(0.751 \text{ M})(125 \text{ mL})}{(125 + 150.) \text{ mL}} = 0.341 \text{ M}$

　　b. $\dfrac{(9.76\%)(155 \text{ g})}{(155 + 125) \text{ g}} = 5.40\%$

　　c. $\dfrac{(18.1 \text{ M})(2.75 \text{ mL})}{(2.75 + 951) \text{ mL}} = 0.0522 \text{ M}$

38. a. $\dfrac{(41.5 \text{ mL})(0.118 \text{ M})(1)}{(0.242 \text{ M})(2)} = 10.1 \text{ mL } H_2SO_4$

　　b. $\dfrac{(27.1 \text{ mL})(0.121 \text{ M})(3)}{(0.242 \text{ M})(2)} = 20.3 \text{ mL } H_2SO_4$

Chapter 16 Equilibrium

2. The nitrogen-nitrogen triple bond in N_2 and the three hydrogen-hydrogen bonds (in the three H_2 molecules) must be broken. Six nitrogen-hydrogen bonds must form (in the two ammonia molecules).

4. The activation energy is the minimum energy two colliding molecules must possess in order for the collision to result in reaction. If molecules do not possess energies equal to or greater than E_a, a collision between these molecules will not result in a reaction.

6. Living cells contain biological catalysts called *enzymes*. Such enzymes are necessary to speed up the complicated biochemical processes that must occur in cells. Such processes would be too slow to sustain life at room temperature if such catalysts were not present.

8. A state of equilibrium is attained when two opposing processes are exactly balanced. The development of a vapor pressure above a liquid in a closed container is an example of a physical equilibrium. Any chemical reaction which appears to "stop" before completion serves as an example of a chemical equilibrium.

10. Chemical equilibrium occurs when two *opposing* chemical reactions reach the *same speed* in a closed system. When a state of chemical equilibrium has been reached, the concentrations of reactants and products present in the system remain *constant* with time, and the reaction appears to "stop." A chemical reaction that reaches a state of equilibrium is indicated by using a double arrow ($\rightleftharpoons$). The points of the double arrow point in opposite directions, indicating that two opposite processes are going on.

12. Although we recognize a state of chemical equilibrium by the fact that the concentrations of reactants and products no longer change with time, the lack of change results from the fact that two *opposing* processes are going on at the same time with the same rate (not because the reaction has truly "stopped"). Further reaction in the forward direction is canceled out by an equal extent of reaction in the reverse direction. The reaction is still proceeding, but the opposite reaction is also proceeding at the same rate.

14. The equilibrium constant is a *ratio* of the concentration of products to the concentration of reactants, with all concentrations measured at equilibrium. Depending on how much reactant a particular experiment was begun with, there may be different absolute amounts of reactants and products present at equilibrium, but the *ratio* will always be the same for a given reaction at a given temperature. For example, the ratios (4/2) and (6/3) are different absolutely in terms of the numbers involved, but each of these ratios has the *value* of 2.

16. a. $K = \dfrac{[NCl_3]^2}{[N_2][Cl_2]^3}$

    b. $K = \dfrac{[HI]^2}{[H_2]^2[I_2]^2}$

Chapter 16   Equilibrium

    c. $K = \dfrac{[N_2H_4]}{[N_2][H_2]^2}$

18. a. $K = \dfrac{[PCl_3]^4}{[P_4][Cl_2]^6}$

    b. $K = \dfrac{[NO]^2}{[N_2][O_2]}$

    c. $K = \dfrac{[CO_2]^3[H_2O]^4}{[C_3H_8][O_2]^5}$

20. $N_2(g) + 3H_2(g) \rightleftharpoons 2NH_3(g)$

$$K = \dfrac{[NH_3]^2}{[N_2][H_2]^3} = \dfrac{[0.34\ M]^2}{[4.9 \times 10^{-4}\ M][2.1 \times 10^{-3}\ M]^3} = 2.5 \times 10^{10}$$

22. $2SO_2(g) + O_2(g) \rightleftharpoons 2SO_3(g)$

$$K = \dfrac{[SO_3]^2}{[SO_2]^2[O_2]} = \dfrac{[4.2 \times 10^{-1}\ M]^2}{[2.3 \times 10^{-3}\ M]^2[3.7 \times 10^{-5}\ M]} = 9.0 \times 10^8$$

24. Equilibrium constants represent ratios of the *concentrations* of products and reactants present at the point of equilibrium. The *concentration* of a pure solid or of a pure liquid is constant and is determined by the density of the solid or liquid. For example, suppose you had a liter of water. Within that liter of water are 55.5 mol of water (the number of moles of water that is contained in one liter of water *does not vary*).

26. a. $K = \dfrac{[SO_2]}{[O_2]}$

    b. $K = \dfrac{[CO_2]^5}{[O_2]^8}$

    c. $K = [NO_2]^2$

28. a. $K = \dfrac{1}{[N_2]}$

    b. $K = [SO_2][HCl]^2$

    c. $K = \dfrac{[H_2]}{[HI]^2}$

30. When an additional amount of one of the reactants is added to an equilibrium system, the system shifts to the right and adjusts so as to use up some of the added reactant. This results in a net *increase* in the amount of product, compared to the equilibrium system before the additional reactant was added. The numerical *value* of the equilibrium constant does *not* change when a reactant is added: the concentrations of

32. If heat is applied to an endothermic reaction (i.e., the temperature is raised), the equilibrium is shifted to the right. More product will be present at equilibrium than if the temperature had not been increased. The value of K increases.

34. a. shifts right (system reacts to get rid of excess SO_2)
    b. shifts right (system reacts to replace missing SO_3)
    c. no change (catalysts do not affect the position of equilibrium)

36. a. no effect (UO_2 is a solid)
    b. no effect (Xe is not involved in the reaction)
    c. shifts to left (if HF attacks glass, it is removed from system, causing reaction to replace the lost HF)
    d. shifts to right
    e. shifts to left (4 mol gas versus 3 mol gas)

38. A decrease in temperature will tend to increase the yield of product. Heat is a product of the reaction; removing heat (decreasing the temperature) favors the forward reaction

40. For an *endo*thermic reaction, an increase in temperature will shift the position of equilibrium to the right (toward products).

42. Heat is a product of the reaction. Removing heat will tend to favor the forward reaction. The reaction is *exo*thermic and should be performed at as low a temperature as possible (consistent with the molecules still having sufficient energy to react).

44. A small equilibrium constant means that the concentration of products is small, compared to the concentration of reactants. The position of equilibrium lies far to the left. Reactions with very small equilibrium constants are generally not very useful as a source of the products, unless Le Châtelier's principle can be applied to shift the position of equilibrium to the point where a sufficient amount of product can be isolated.

46. $K = \dfrac{[NCl_3]^2}{[N_2][Cl_2]^3} = \dfrac{[1.9 \times 10^{-1}]^2}{[1.4 \times 10^{-3} \, M][4.3 \times 10^{-4} \, M]^3} = 3.2 \times 10^{11}$

48. $K = [CO_2] = [2.1 \times 10^{-3} \, M] = 2.1 \times 10^{-3}$

50. $K = \dfrac{[H_2]^2[O_2]}{[H_2O]^2}$

$$2.4 \times 10^{-3} = \frac{[1.9 \times 10^{-2}]^2 [O_2]}{[1.1 \times 10^{-1}]^2}$$

$$[O_2] = \frac{(2.4 \times 10^{-3})(1.1 \times 10^{-1})^2}{(1.9 \times 10^{-2})^2} = 8.0 \times 10^{-2} \text{ M}$$

52. $K = \dfrac{[N_2O_4]}{[NO_2]^2}$

$$1.2 \times 10^4 = \frac{(0.45 \text{ M})}{[NO_2]^2}$$

$$[NO_2]^2 = \frac{(0.45)}{1.2 \times 10^4} = 3.75 \times 10^{-5}$$

$[NO_2] = 6.1 \times 10^{-3}$ M

54. solubility product, K_{sp}

56. Stirring or grinding the solute increases the speed with which the solute dissolves, but the ultimate *amount* of solute that dissolves is fixed by the equilibrium constant for the dissolving process, K_{sp}.

58. a. $SnS_2(s) \rightleftharpoons Sn^{4+}(aq) + 2S^{2-}(aq)$
$K_{sp} = [Sn^{4+}][S^{2-}]^2$
b. $AlPO_4(s) \rightleftharpoons Al^{3+}(aq) + PO_4^{3-}(aq)$
$K_{sp} = [Al^{3+}][PO_4^{3-}]$
c. $Ni(OH)_2(s) \rightleftharpoons Ni^{2+}(aq) + 2OH^-(aq)$
$K_{sp} = [Ni^{2+}][OH^-]^2$
d. $Hg_2Cl_2(s) \rightleftharpoons Hg_2^{2+}(aq) + 2Cl^-(aq)$
$K_{sp} = [Hg_2^{2+}][Cl^-]^2$

60. $ZnCO_3(s) \rightleftharpoons Zn^{2+}(aq) + CO_3^{2-}(aq)$
$K_{sp} = [Zn^{2+}][CO_3^{2-}]$
If 1.7×10^{-5} mol of $ZnCO_3$ dissolve per liter, then the concentrations of the two ions produced will each also be 1.7×10^{-5} M.
$K_{sp} = [1.7 \times 10^{-5} \text{ M}][1.7 \times 10^{-5} \text{ M}] = 2.9 \times 10^{-10}$.

62. $CuCrO_4(s) \rightleftharpoons Cu^{2+}(aq) + CrO_4^{2-}(aq)$
$K_{sp} = [Cu^{2+}][CrO_4^{2-}]$
Molar mass of $CuCrO_4$ = 179.55 g

$$1.1 \times 10^{-5} \text{ g CuCrO}_4 \times \frac{1 \text{ mol}}{179.55 \text{ g}} = 6.1 \times 10^{-8} \text{ mol}$$

If 6.1×10^{-8} mol of $CuCrO_4$ dissolve per liter, then the concentrations of $Cu^{2+}(aq)$ and $CrO_4^{2-}(aq)$ are also each 6.1×10^{-8} M.

$K_{sp} = (6.1 \times 10^{-8} M)(6.1 \times 10^{-8} M) = 3.8 \times 10^{-15}$

64. $SrCO_3(s) \rightleftharpoons Sr^{2+}(aq) + CO_3^{2-}(aq)$

    $K_{sp} = [Sr^{2+}][CO_3^{2-}] = 1.1 \times 10^{-10}$

    Let x represent the number of moles of $SrCO_3(s)$ that dissolve per liter, then $[Sr^{2+}] = x$ and $[CO_3^{2-}] = x$ also from the stoichiometry of the reaction; then

    $K_{sp} = [x][x] = x^2 = 1.1 \times 10^{-10}$

    $x = [SrCO_3] = 1.0 \times 10^{-5}$ M

66. $PbCrO_4(s) \rightleftharpoons Pb^{2+}(aq) + CrO_4^{2-}(aq)$

    $K_{sp} = [Pb^{2+}][CrO_4^{2-}] = 2.8 \times 10^{-13}$

    Let x represent the number of moles of $PbCrO_4(s)$ that dissolve per liter. Then $[Pb^{2+}] = x$ and $[CrO_4^{2-}] = x$ also from the stoichiometry of the reaction; then

    $K_{sp} = [x][x] = x^2 = 2.8 \times 10^{-13}$

    $x = [PbCrO_4] = 5.3 \times 10^{-7}$ M.

    Molar mass of $PbCrO_4 = 323.2$ g

    5.3×10^{-7} mol/L $\times 323.2$ g/mol $= 1.7 \times 10^{-4}$ g/L

68. $PbCl_2(s) \rightleftharpoons Pb^{2+}(aq) + 2Cl^-(aq)$

    $K_{sp} = [Pb^{2+}][Cl^-]^2$

    If $PbCl_2$ dissolves to the extent of 3.6×10^{-2} M, then

    $[Pb^{2+}] = 3.6 \times 10^{-2}$ M and

    $[Cl^-] = 2 \times (3.6 \times 10^{-2}) = 7.2 \times 10^{-2}$ M

    $K_{sp} = (3.6 \times 10^{-2} M)(7.2 \times 10^{-2} M)^2 = 1.9 \times 10^{-4}$

    molar mass $PbCl_2 = 278.1$ g

    $3.6 \times 10^{-2} \dfrac{\text{mol}}{1 \text{ L}} \times \dfrac{278.1 \text{ g}}{1 \text{ mol}} = 10.$ g/L

70. $Fe(OH)_3(s) \rightleftharpoons Fe^{3+}(aq) + 3OH^-(aq)$

    $K_{sp} = [Fe^{3+}][OH^-]^3 = 4 \times 10^{-38}$

    Let x represent the number of moles of $Fe(OH)_3$ that dissolve per liter; then $[Fe^{3+}] = x$.

    The amount of hydroxide ion that would be produced by the dissolving of the $Fe(OH)_3$ would then be $3x$, but pure water itself contains hydroxide ion at the concentration of 1.0×10^{-7} M (see Chapter 17). The total concentration of hydroxide ion is then $[OH^-] = (3x + 1.0 \times 10^{-7})$. Since x must be a very small number [since Fe(OH)$_3$ is not very soluble], we can save ourselves a lot of arithmetic if we use the approximation that $(3x + 1.0 \times 10^{-7}$ M$) = 1.0 \times 10^{-7}$

$K_{sp} = [x][1.0 \times 10^{-7}]^3 = 4 \times 10^{-38}$

$x = 4 \times 10^{-17}$ M

molar mass Fe(OH)$_3$ = 106.9 g

$$\frac{4 \times 10^{-17} \text{ mol}}{1.00 \text{ L}} \times \frac{106.9 \text{ g}}{1 \text{ mol}} = 4 \times 10^{-15} \text{ g/L}$$

72. temperature

74. catalyst

76. constant

78. dynamic

80. heterogeneous

82. position

84. exothermic

86. An equilibrium reaction may come to many *positions* of equilibrium, but at each possible position of equilibrium, the numerical value of the equilibrium constant is fulfilled. If different amounts of reactant are taken in different experiments, the *absolute amounts* of reactant and product present at the point of equilibrium reached will differ from one experiment to another, but the *ratio* that defines the equilibrium constant will be the same.

88. $PCl_5(g) \rightleftharpoons PCl_3(g) + Cl_2(g)$

$$K = \frac{[PCl_3][Cl_2]}{[PCl_5]} = 4.5 \times 10^{-3}$$

The concentration of PCl$_5$ is to be twice the concentration of PCl$_3$:

$[PCl_5] = 2 \times [PCl_3]$

$$K = \frac{[PCl_3][Cl_2]}{2 \times [PCl_3]} = 4.5 \times 10^{-3}$$

$$K = \frac{[Cl_2]}{2} = 4.5 \times 10^{-3} \quad \text{and} \quad [Cl_2] = 9.0 \times 10^{-3} \text{ M}$$

Chapter 16   Equilibrium   213

90. $Ag_2CrO_4(s) \rightleftharpoons 2Ag^+(aq) + CrO_4^{2-}(aq)$

$K_{sp} = [Ag^+]^2[CrO_4^{2-}] = 9.1 \times 10^{-12}$

let x represent the number of moles of Ag_2CrO_4 that dissolve per liter; then

$[Ag^+] = 2x$ and $[CrO_4^{2-}] = x$

$K_{sp} = [2x]^2[x] = 9.1 \times 10^{-12}$

$4x^3 = 9.1 \times 10^{-12}$

$x^3 = 2.275 \times 10^{-12}$

$x = 1.3 \times 10^{-4}$ M

molar mass $Ag_2CrO_4 = 331.8$ g

$$\frac{1.3 \times 10^{-4} \text{ mol}}{1.00 \text{ L}} \times \frac{331.8 \text{ g}}{1 \text{ mol}} = 0.043 \text{ g/L}$$

92. Although a small solubility product generally implies a small solubility, comparisons of solubility based directly on K_{sp} values are only valid if the salts produce the same numbers of positive and negative ions per formula when they dissolve. For example, one can compare the solubilities of $AgCl(s)$ and $NiS(s)$ directly using K_{sp}, since each salt produces one positive and one negative ion per formula when dissolved. One could not directly compare $AgCl(s)$ with a salt such as $Ca_3(PO_4)_2$, however.

94. At higher temperatures, the average kinetic energy of the reactant molecules is larger. At higher temperatures, the probability that a collision between molecules will be energetic enough for reaction to take place is larger. On a molecular basis, a higher temperature means a given molecule will be moving faster.

96. a. $K = \dfrac{[HBr]^2}{[H_2][Br_2]}$

b. $K = \dfrac{[H_2S]^2}{[H_2]^2[S_2]}$

c. $K = \dfrac{[HCN]^2}{[H_2][C_2N_2]}$

98. $K = \dfrac{[Br]^2}{[Br_2]} = \dfrac{(0.034 \text{ M})^2}{(0.97 \text{ M})} = 1.2 \times 10^{-3}$

100. a. $K = \dfrac{1}{[O_2]^3}$

b. $K = \dfrac{1}{[NH_3][HCl]}$

c. $K = \dfrac{1}{[O_2]}$

102. An *exothermic* reaction is one which liberates heat energy. Increasing the temperature (adding heat) for such a reaction is fighting against the reaction's own tendency to liberate heat. The net effect of raising the temperature will be a shift to the left to decrease the amount of product. If it is desired to increase the amount of product in an exothermic reaction, heat must be *removed* from the system. Changing the temperature *does* change the numerical value of the equilibrium constant for a reaction.

104. The reaction is *exothermic* as written. An increase in temperature (addition of heat) will shift the reaction to the left (toward reactants).

106. $K = \dfrac{[NH_3]^2}{[N_2][H_2]^3}$

$1.3 \times 10^{-2} = \dfrac{[NH_3]}{(0.10\ M)(0.10\ M)^3}$

$[NH_3]^2 = 1.3 \times 10^{-2} \times (0.10) \times (0.10)^3 = 1.3 \times 10^{-6}$

$[NH_3] = 1.1 \times 10^{-3}\ M$

108. a. $MgCO_3(s) \rightleftharpoons Mg^{2+}(aq) + CO_3^{2-}(aq)$
$K_{sp} = [Mg^{2+}][CO_3^{2-}]$
b. $Al_2(SO_4)_3(s) \rightleftharpoons 2Al^{3+}(aq) + 3SO_4^{2-}(aq)$
$K_{sp} = [Al^{3+}]^2[SO_4^{2-}]^3$
c. $Ca(OH)_2(s) \rightleftharpoons Ca^{2+}(aq) + 2OH^-(aq)$
$K_{sp} = [Ca^{2+}][OH^-]^2$
d. $Ag_2SO_4(s) \rightleftharpoons 2Ag^+(aq) + SO_4^{2-}(aq)$
$K_{sp} = [Ag^+]^2[SO_4^{2-}]$

110. molar mass AgCl = 143.4 g

9.0×10^{-4} g AgCl/L $\times \dfrac{1\ mol\ AgCl}{143.4\ g\ AgCl} = 6.28 \times 10^{-6}$ mol AgCl/L

$AgCl(s) \rightleftharpoons Ag^+(aq) + Cl^-(aq)$

$K_{sp} = [Ag^+][Cl^-] = (6.28 \times 10^{-6}\ M)(6.28 \times 10^{-6}\ M) = 3.9 \times 10^{-11}$

112. molar mass Ni(OH)$_2$ = 92.71 g

$$\frac{0.14 \text{ g Ni(OH)}_2}{1.00 \text{ L}} \times \frac{1 \text{ mol Ni(OH)}_2}{92.71 \text{ g Ni(OH)}_2} = 1.510 \times 10^{-3} \text{ M}$$

Ni(OH)$_2$(s) ⇌ Ni^{2+}(aq) + 2OH$^-$(aq)

K_{sp} = [Ni^{2+}][OH$^-$]^2

If 1.510×10^{-3} M of Ni(OH)$_2$ dissolves, then

[Ni^{2+}] = 1.510×10^{-3} M and

[OH$^-$] = 2 × (1.510×10^{-3} M) = 3.020×10^{-3} M

K_{sp} = (1.510×10^{-3} M)(3.020×10^{-3} M)^2 = 1.4×10^{-8}

Chapter 17   Acids and Bases

2. In the Arrhenius definition, an acid is a substance which produces hydrogen ions (H$^+$) when dissolved in water, whereas a base is a substance which produces hydroxide ions (OH$^-$) in aqueous solution. These definitions proved to be too restrictive since the only base permitted was hydroxide ion, and the only solvent permitted was water.

4. A conjugate acid-base pair differ from each other by one proton (one hydrogen ion, H$^+$). For example, CH$_3$COOH (acetic acid), differs from its conjugate base, CH$_3$COO$^-$ (acetate ion), by a single H$^+$ ion.

$$CH_3COOH(aq) \rightarrow CH_3COO^-(aq) + H^+(aq)$$

6. When an acid is dissolved in water, the hydronium ion (H$_3$O$^+$) is formed. The hydronium ion is the conjugate *acid* of water (H$_2$O).

8. a. HSO$_4^-$ and SO$_4^{2-}$ represent a conjugate acid-base pair (HSO$_4^-$ is the acid, SO$_4^{2-}$ is the base; they differ by one proton).

   b. HBr and BrO$^-$ are not a conjugate acid-base pair (Br$^-$ is the conjugate base of HBr; BrO$^-$ is the conjugate base of HBrO).

   c. H$_2$PO$_4^-$ and PO$_4^{3-}$ are not a conjugate acid-base pair; they differ by two protons (H$_2$PO$_4^-$ is the conjugate acid of HPO$_4^{2-}$ and also the conjugate base of H$_3$PO$_4$; HPO$_4^{2-}$ is the conjugate acid of PO$_4^{3-}$).

   d. HNO$_3$ and NO$_2^-$ are not a conjugate acid-base pair; they differ by an oxygen atom as well as a proton (NO$_3^-$ is the conjugate base of HNO$_3$; NO$_2^-$ is the conjugate base of HNO$_2$).

10. a. HSO$_4^-$(acid), SO$_4^{2-}$(base); H$_2$O(base), H$_3$O$^+$(acid)

    b. CH$_3$NH$_2$(base), CH$_3$NH$_3^+$(acid); HCl(acid), Cl$^-$(base)

    c. CH$_3$COO$^-$(base), CH$_3$COOH(acid); H$_2$O(acid), OH$^-$(base)

12. The conjugate *acid* of the species indicated would have *one additional proton*:

    a. HSO$_4^-$
    b. HPO$_4^{2-}$
    c. CH$_3$NH$_3^+$
    d. HF

14. The conjugate *bases* of the species indicated would have *one less proton*:

    a. NO$_3^-$
    b. PO$_4^{3-}$
    c. H$_2$O
    d. HCO$_3^-$

16. a. $H_3PO_4 + H_2O \rightleftharpoons H_2PO_4^- + H_3O^+$
    b. $H_2PO_4^- + H_2O \rightleftharpoons HPO_4^{2-} + H_3O^+$
    c. $NH_4^+ + H_2O \rightleftharpoons NH_3 + H_3O^+$
    d. $H_2S + H_2O \rightleftharpoons HS^- + H_3O^+$

18. To say that an acid is *weak* in aqueous solution means that the acid does not easily transfer protons to water (and does not fully ionize). If an acid does not lose protons easily, then the acid's anion must be a strong attractor of protons (good at holding on to protons).

20. A strong acid is one which loses its protons easily and fully ionizes in water; this means that the acid's conjugate base must be poor at attracting and holding on to protons, and is therefore a relatively weak base. A weak acid is one which resists loss of its protons and does not ionize well in water; this means that the acid's conjugate base attracts and holds onto protons tightly and is a relatively strong base.

22. H_2SO_4 (sulfuric): $H_2SO_4 + H_2O \rightarrow HSO_4^- + H_3O^+$

    HCl (hydrochloric): $HCl + H_2O \rightarrow Cl^- + H_3O^+$

    HNO_3 (nitric): $HNO_3 + H_2O \rightarrow NO_3^- + H_3O^+$

    $HClO_4$ (perchloric): $HClO_4 + H_2O \rightarrow ClO_4^- + H_3O^+$

24. oxyacids: $HClO_4$, HNO_3, H_2SO_4, CH_3COOH, etc.

    non-oxyacids: HCl, HBr, HF, HI, HCN, etc.

26. Bases that are *weak* have relatively strong conjugate acids:
    a. SO_4^{2-} is a moderately weak base; HSO_4^- is a moderately strong acid
    b. Br^- is a very weak base; HBr is a strong acid
    c. CN^- is a fairly strong base; HCN is a fairly weak acid
    d. CH_3COO^- is a fairly strong base; CH_3COOH is a weak acid

28. For example, HCO_3^- can behave as an acid if it reacts with something that more strongly gains protons than does HCO_3^- itself. For example, HCO_3^- would behave as an acid when reacting with hydroxide ion (a much stronger base).

    $HCO_3^-(aq) + OH^-(aq) \rightarrow CO_3^{2-}(aq) + H_2O(l)$

    On the other hand, HCO_3^- would behave as a base when reacted with something that more readily loses protons than does HCO_3^- itself. For example, HCO_3^- would behave as a base when reacting with hydrochloric acid (a much stronger acid).

    $HCO_3^-(aq) + HCl(aq) \rightarrow H_2CO_3(aq) + Cl^-(aq)$

218 Chapter 17 Acids and Bases

For $H_2PO_4^-$, similar equations can be written:

$$H_2PO_4^-(aq) + OH^-(aq) \rightarrow HPO_4^{2-}(aq) + H_2O(l)$$

$$H_2PO_4^-(aq) + H_3O^+(aq) \rightarrow H_3PO_4(aq) + H_2O(l)$$

30. The hydrogen ion concentration and the hydroxide ion concentration of water are *not* independent of each other: they are related by the equilibrium

$$H_2O(l) \rightleftharpoons H^+(aq) + OH^-(aq)$$

for which $K_w = [H^+][OH^-] = 1.0 \times 10^{-14}$ at 25°C.

If the concentration of one of these ions is increased by addition of a reagent producing H^+ or OH^-, then the concentration of the complementary ion will have to decrease so that the value of K_w will hold true. So if an acid is added to a solution, the concentration of hydroxide ion in the solution will decrease to a lower value. Similarly, if a base is added to a solution, then the concentration of hydrogen ion will have to decrease to a lower value.

32. $K_w = [H^+][OH^-] = 1.0 \times 10^{-14}$ at 25°C

$$[H^+] = \frac{1.0 \times 10^{-14}}{[OH^-]}$$

a. $[H^+] = \dfrac{1.0 \times 10^{-14}}{3.99 \times 10^{-5} \, M} = 2.5 \times 10^{-10}$ M; solution is basic

b. $[H^+] = \dfrac{1.0 \times 10^{-14}}{2.91 \times 10^{-9} \, M} = 3.4 \times 10^{-6}$ M; solution is acidic

c. $[H^+] = \dfrac{1.0 \times 10^{-14}}{7.23 \times 10^{-2} \, M} = 1.4 \times 10^{-13}$ M; solution is basic

d. $[H^+] = \dfrac{1.0 \times 10^{-14}}{9.11 \times 10^{-7} \, M} = 1.1 \times 10^{-8}$ M; solution is basic

34. $[OH^-] = \dfrac{1.0 \times 10^{-14}}{[H^+]}$

a. $[OH^-] = \dfrac{1.0 \times 10^{-14}}{1.27 \times 10^{-3} \, M} = 7.9 \times 10^{-12}$ M; solution is acidic

b. $[OH^-] = \dfrac{1.0 \times 10^{-14}}{4.49 \times 10^{-8} \, M} = 2.2 \times 10^{-7}$ M; solution is basic

c. $[OH^-] = \dfrac{1.0 \times 10^{-14}}{3.87 \times 10^{-5} \, M} = 2.6 \times 10^{-10}$ M; solution is acidic

d. $[OH^-] = \dfrac{1.0 \times 10^{-14}}{8.16 \times 10^{-11} \, M} = 1.2 \times 10^{-4}$ M; solution is basic

36. a. $[OH^-] = 0.10\ M$ is more basic
    b. $[H^+] = 0.00100\ M$ is more basic
    c. $[H^+] = 1.15 \times 10^{-7}\ M$ is more basic

38. household ammonia (pH 12); blood (pH 7-8); milk (pH 6-7); vinegar (pH 3); lemon juice (pH 2-3); stomach acid (pH 2)

40. The pH of a solution is defined as the *negative* of the logarithm of the hydrogen ion concentration, pH = $-\log[H^+]$. Mathematically, the *negative sign* in the definition causes the pH to *decrease* as the hydrogen ion concentration *increases*.

42. pH = $-\log[H^+]$
    a. pH = $-\log[0.0010\ M]$ = 3.000; solution is acidic
    b. pH = $-\log[2.19 \times 10^{-4}\ M]$ = 3.660; solution is acidic
    c. pH = $-\log[9.18 \times 10^{-11}\ M]$ = 10.037; solution is basic
    d. pH = $-\log[4.71 \times 10^{-7}\ M]$ = 6.327; solution is acidic

44. pOH = $-\log[OH^-]$ pH = 14 − pOH
    a. pOH = $-\log[4.22 \times 10^{-3}\ M]$ = 2.375
       pH = 14 − 2.375 = 11.625 = 11.63; solution is basic
    b. pOH = $-\log[1.99 \times 10^{-7}\ M]$ = 6.701
       pH = 14 − 6.701 = 7.299 = 7.30; solution is basic
    c. pOH = $-\log[0.000100\ M]$ = 4.000
       pH = 14 − 4.000 = 10.00; solution is basic
    d. pOH = $-\log[2.47 \times 10^{-11}\ M]$ = 10.607
       pH = 14 − 10.607 = 3.393 = 3.39; solution is acidic

46. pOH = 14 − pH
    a. pOH = 14 − 6.49 = 7.51; solution is acidic
    b. pOH = 14 − 1.93 = 12.07; solution is acidic
    c. pOH = 14 − 11.21 = 2.79; solution is basic
    d. pOH = 14 − 7.00 = 7.00; solution is neutral

48. a. $[H^+] = 1.00 \times 10^{-7}\ M$

    $[OH^-] = \dfrac{1.0 \times 10^{-14}}{1.00 \times 10^{-7}\ M} = 1.0 \times 10^{-7}\ M$

    pH = $-\log[1.0 \times 10^{-7}\ M]$ = 7.00
    pOH = 14 − 7.00 = 7.00

220 Chapter 17 Acids and Bases

b. $[OH^-] = 4.39 \times 10^{-5}$ M

$[H^+] = \dfrac{1.0 \times 10^{-14}}{4.39 \times 10^{-5} \text{ M}} = 2.28 \times 10^{-10}$ M $= 2.3 \times 10^{-10}$ M

pH $= -\log[2.28 \times 10^{-10}$ M$] = 9.64$

pOH $= 14 - 9.64 = 4.36$

c. $[H^+] = 4.29 \times 10^{-11}$ M

$[OH^-] = \dfrac{1.0 \times 10^{-14}}{4.29 \times 10^{-11} \text{ M}} = 2.33 \times 10^{-4}$ M $= 2.3 \times 10^{-4}$ M

pH $= -\log[4.29 \times 10^{-11}$ M$] = 10.37$

pOH $= 14 - 10.368 = 3.63$

d. $[OH^-] = 7.36 \times 10^{-2}$ M

$[H^+] = \dfrac{1.0 \times 10^{-14}}{7.36 \times 10^{-2} \text{ M}} = 1.36 \times 10^{-13}$ M $= 1.4 \times 10^{-13}$ M

pH $= -\log[1.36 \times 10^{-13}$ M$] = 12.87$

pOH $= 14 - 12.87 = 1.13$

50. $[H^+] = \{inv\}\{log\}[-pH]$

a. $[H^+] = \{inv\}\{log\}[-4.11] = 7.8 \times 10^{-5}$ M
b. $[H^+] = \{inv\}\{log\}[-9.54] = 2.9 \times 10^{-10}$ M
c. $[H^+] = \{inv\}\{log\}[-7.00] = 1.0 \times 10^{-7}$ M
d. $[H^+] = \{inv\}\{log\}[-13.21] = 6.2 \times 10^{-14}$ M

52. a. pH $= 14 - 3.91 = 10.09$

$[H^+] = \{inv\}\{log\}[-10.09] = 8.1 \times 10^{-11}$ M

b. pH $= 14 - 12.56 = 1.44$

$[H^+] = \{inv\}\{log\}[-1.44] = 3.6 \times 10^{-2}$ M

c. pH $= 14 - 1.15 = 12.85$

$[H^+] = \{inv\}\{log\}[-12.85] = 1.4 \times 10^{-13}$ M

d. pH $= 14 - 8.77 = 5.23$

$[H^+] = \{inv\}\{log\}[-5.23] = 5.9 \times 10^{-6}$ M

54. a. $[H^+] = \{inv\}\{log\}[-4.38] = 4.2 \times 10^{-5}$ M

b. pH $= 14 - 4.38 = 9.62$

$[H^+] = \{inv\}\{log\}[-9.62] = 2.4 \times 10^{-10}$ M

c. $[H^+] = \{inv\}\{log\}[-7.21] = 6.2 \times 10^{-8}$ M

d. pH $= 14 - 5.39 = 8.61$

$[H^+] = \{inv\}\{log\}[-8.61] = 2.5 \times 10^{-9}$ M

56. The solution contains water molecules, H_3O^+ ions (protons), and NO_3^- ions. Because HNO_3 is a strong acid, which is completely ionized in water, there are no HNO_3 molecules present.

58. a. HCl is a strong acid and completely ionized so

   $[H^+] = 0.00010\ M$ and pH = 4.00

   b. HNO_3 is a strong acid and completely ionized so

   $[H^+] = 0.0050\ M$ and pH = 2.30

   c. $HClO_4$ is a strong acid and completely ionized so

   $[H^+] = 4.21 \times 10^{-5}\ M$ and pH = 4.376

   d. HNO_3 is a strong acid and completely ionized so

   $[H^+] = 6.33 \times 10^{-3}\ M$ and pH = 2.199

60. A buffered solution consists of a mixture of a weak acid and its conjugate base; one example of a buffered solution is a mixture of acetic acid (CH_3COOH) and sodium acetate ($NaCH_3COO$).

62. The weak acid component of a buffered solution is capable of reacting with added strong base. For example, using the buffered solution given as an example in Question 60, acetic acid would consume added sodium hydroxide as follows:

   $CH_3COOH(aq) + NaOH(aq) \rightarrow NaCH_3COO(aq) + H_2O(l)$

   Acetic acid *neutralizes* the added NaOH and prevents it from having much effect on the overall pH of the solution.

64. added NaOH: $HF + OH^- \rightarrow F^- + H_2O$
    added HCl: $F^- + H_3O^+ \rightarrow HF + H_2O$
    added NaOH: $NH_4^+ + OH^- \rightarrow NH_3 + H_2O$
    added HCl: $NH_3 + H_3O^+ \rightarrow NH_4^+ + H_2O$

66. a. NaOH is completely ionized, so $[OH^-] = 0.10\ M$

   pOH = $-\log[0.10] = 1.00$

   pH = 14 − 1.00 = 13.00

   b. KOH is completely ionized, so $[OH^-] = 2.0 \times 10^{-4}\ M$

   pOH = $-\log[2.0 \times 10^{-4}] = 3.70$

   pH = 14 − 3.70 = 10.30

   c. CsOH is completely ionized, so $[OH^-] = 6.2 \times 10^{-3}\ M$

   pOH = $-\log[6.2 \times 10^{-3}] = 2.21$

   pH = 14 − 2.21 = 11.79

222   Chapter 17   Acids and Bases

    d.   NaOH is completely ionized, so [OH⁻] = 0.0001 M

          pOH = −log[0.0001] = 4.0

          pH = 14 − 4.0 = 10.0

68. *a*, *b*, and *d* represent basic solutions; *c* represents an *acidic* solution because there is less hydroxide ion than hydrogen ion.

70. *a*, *c*, and *e* represent strong acids; *b* and *d* are typical *weak* acids.

72. Ordinarily in calculating the pH of strong acid solutions, the major contribution to the concentration of hydrogen ion present is from the dissolved strong acid; we ordinarily neglect the small amount of hydrogen ion present in such solutions due to the ionization of water. With 1.0×10^{-7} M HCl solution, however, the amount of hydrogen ion present due to the ionization of *water* is *comparable* to that present due to the addition of *acid* (HCl) and must be considered in the calculation of pH.

74. accepts

76. base

78. carboxyl (-COOH)

80. 1.0×10^{-14}

82. larger

84. pH

86. weak acid

88.   a. H_2O and OH^- represent a conjugate acid-base pair (H_2O is the acid, having one more proton than the base, OH^-).

     b. H_2SO_4 and SO_4^{2-} are *not* a conjugate acid-base pair (they differ by *two* protons). The conjugate base of H_2SO_4 is HSO_4^-; the conjugate acid of SO_4^{2-} is also HSO_4^-.

     c. H_3PO_4 and $H_2PO_4^-$ represent a conjugate acid-base pair (H_3PO_4 is the acid, having one more proton than the base $H_2PO_4^-$).

     d. $HC_2H_3O_2$ and $C_2H_3O_2^-$ represent a conjugate acid-base pair ($HC_2H_3O_2$ is the acid, having one more proton than the base $C_2H_3O_2^-$).

90. The conjugate *acid* of the species indicated would have *one additional proton*:

    a. NH_4^+ c. H_3PO_4

    b. CH_3COOH d. $H_2C_2O_4$

92. When an acid ionizes in water, a proton is released to the water as an H_3O^+ ion:

    a. $CH_3CH_2COOH + H_2O \rightleftharpoons CH_3CH_2COO^- + H_3O^+$

    b. $NH_4^+ + H_2O \rightleftharpoons NH_3 + H_3O^+$

    c. $H_2SO_4 + H_2O \rightleftharpoons HSO_4^- + H_3O^+$

    d. $H_3PO_4 + H_2O \rightleftharpoons H_2PO_4^- + H_3O^+$

94. $K_w = [H^+][OH^-] = 1.0 \times 10^{-14}$ at 25°C

$$[H^+] = \frac{1.0 \times 10^{-14}}{[OH^-]}$$

    a. $[H^+] = \dfrac{1.0 \times 10^{-14}}{4.22 \times 10^{-3} \, M} = 2.4 \times 10^{-12} \, M$; solution is basic

    b. $[H^+] = \dfrac{1.0 \times 10^{-14}}{1.01 \times 10^{-13} \, M} = 9.9 \times 10^{-2} \, M$; solution is acidic

    c. $[H^+] = \dfrac{1.0 \times 10^{-14}}{3.05 \times 10^{-7} \, M} = 3.3 \times 10^{-8} \, M$; solution is basic

    d. $[H^+] = \dfrac{1.0 \times 10^{-14}}{6.02 \times 10^{-6} \, M} = 1.7 \times 10^{-9} \, M$; solution is basic

96.   a. $[OH^-] = 0.0000032 \, M$ is more basic

    b. $[OH^-] = 1.54 \times 10^{-8} \, M$ is more basic

    c. $[OH^-] = 4.02 \times 10^{-7} \, M$ is more basic

98. $pOH = -\log[OH^-]$ $pH = 14 - pOH$

    a. $pOH = -\log[1.4 \times 10^{-6} \, M] = 5.85$

        $pH = 14 - 5.85 = 8.15$; solution is basic

    b. $pOH = -\log[9.35 \times 10^{-9} \, M] = 8.029 = 8.03$

        $pH = 14 - 8.029 = 5.971 = 5.97$; solution is acidic

    c. $pOH = -\log[2.21 \times 10^{-1} \, M] = 0.656 = 0.66$

        $pH = 14 - 0.656 = 13.344 = 13.34$; solution is basic

    d. $pOH = -\log[7.98 \times 10^{-12} \, M] = 11.098 = 11.10$

        $pH = 14 - 11.098 = 2.902 = 2.90$; solution is acidic

100.   a. $[OH^-] = \dfrac{1.0 \times 10^{-14}}{5.72 \times 10^{-4} \, M} = 1.75 \times 10^{-11} \, M = 1.8 \times 10^{-11} \, M$

        $pOH = -\log[1.75 \times 10^{-11} \, M] = 10.76$

        $pH = 14 - 10.76 = 3.24$

b. $[H^+] = \dfrac{1.0 \times 10^{-14}}{8.91 \times 10^{-5} \, M} = 1.12 \times 10^{-10} \, M = 1.1 \times 10^{-10} \, M$

$pH = -\log[1.12 \times 10^{-10} \, M] = 9.95$

$pOH = 14 - 9.95 = 4.05$

c. $[OH^-] = \dfrac{1.0 \times 10^{-14}}{2.87 \times 10^{-12} \, M} = 3.48 \times 10^{-3} \, M = 3.5 \times 10^{-3} \, M$

$pOH = -\log[3.48 \times 10^{-3} \, M] = 2.46$

$pH = 14 - 2.46 = 11.54$

d. $[H^+] = \dfrac{1.0 \times 10^{-14}}{7.22 \times 10^{-8} \, M} = 1.39 \times 10^{-7} \, M = 1.4 \times 10^{-7} \, M$

$pH = -\log[1.39 \times 10^{-7} \, M] = 6.86$

$pOH = 14 - 6.86 = 7.14$

102. $pH = 14 - pOH$ $[H^+] = \{inv\}\{log\}[-pH]$

a. $pH = 14 - 7.95 = 6.05$

$[H^+] = \{inv\}\{log\}[-6.05] = 8.9 \times 10^{-7} \, M$

b. $pH = 14 - 14.00 = 0.00$

$[H^+] = \{inv\}\{log\}[0.00] = 1.0 \, M$

c. $pH = 14 - 5.00 = 9.00$

$[H^+] = \{inv\}\{log\}[-9.00] = 1.0 \times 10^{-9} \, M$

d. $pH = 14 - 3.95 = 10.05$

$[H^+] = \{inv\}\{log\}[-10.05] = 8.9 \times 10^{-11} \, M$

104. a. $HClO_4$ is a strong acid and completely ionized so
$[H^+] = 1.4 \times 10^{-3} \, M$ and $pH = 2.85$

b. HCl is a strong acid and completely ionized so
$[H^+] = 3.0 \times 10^{-5} \, M$ and $pH = 4.52$

c. HNO_3 is a strong acid and completely ionized so
$[H^+] = 5.0 \times 10^{-2} \, M$ and $pH = 1.30$

d. HCl is a strong acid and completely ionized so
$[H^+] = 0.0010 \, M$ and $pH = 3.00$

Cumulative Review: Chapters 16 and 17

2. A graph illustrating the activation energy barrier for a reaction is given as Figure 16.4 in the text. The activation energy for a reaction represents the minimum energy the reactant molecules must possess for a reaction to occur when the molecules collide. Although an increase in temperature does not change the activation energy for a reaction itself, at higher temperatures, the reactant molecules have a higher average kinetic energy, and a larger fraction of reaction molecules will possess enough energy for a collision to be effective: you will remember that in the kinetic-molecular theory, temperature was a direct measure of average kinetic energy. A catalyst speeds up a reaction by providing an alternate mechanism or pathway by which the reaction can occur, with such a mechanism or pathway having a lower activation energy than the original pathway.

4. The expression for the equilibrium constant for a reaction has as its numerator the concentrations of the products (raised to the powers of their stoichiometric coefficients in the balanced chemical equation for the reaction), and as its denominator the concentrations of the reactants (also raised to the powers of their stoichiometric coefficients). In general terms, for a reaction

$$aA + bB = cC + dD$$

the equilibrium constant expression has the form

$$K = \frac{[C]^c[D]^d}{[A]^a[B]^b}$$

in which square brackets [] indicate concentration. For example, here are three simple reactions and the expression for their equilibrium constants:

$N_2(g) + O_2(g) \rightleftharpoons 2NO(g)$ $K = [NO]^2/[N_2][O_2]$

$2SO_2(g) + O_2(g) \rightleftharpoons 2SO_3(g)$ $K = [SO_3]^2/[SO_2]^2[O_2]$

$N_2(g) + 3H_2(g) \rightleftharpoons 2NH_3(g)$ $K = [NH_3]^2/[N_2][H_2]^3$

6. In a homogeneous equilibrium, all the reactants and products are in the same phase and have the same physical state (solid, liquid, or gas). For a heterogeneous equilibrium, however, one or more of the reactant or products exists in a phase or physical state different from the other substances. When we have a heterogeneous equilibrium, the concentrations of solids and pure liquids are left out of the expression for the equilibrium constant for the reaction: the concentration of a solid or pure liquid is constant.

$C(s) + O_2(g) \rightleftharpoons CO_2(g)$ heterogeneous (solid, gases)

$K = [CO_2]/[O_2]$

$2CO(g) + O_2(g) \rightleftharpoons 2CO_2(g)$ homogeneous (all gases)

$K = [CO_2]^2/[CO]^2[O_2]$

8. When a slightly soluble salt is placed in water, ions begin to leave the crystals of salt and to enter the solvent to form a solution. As the concentration of ions in solution increases, eventually ions from the solution are attracted to and rejoin the crystals of undissolved salt. Eventually things get to the point that every time ions leave the crystals to enter the solution in one place in the system, somewhere else ions are leaving the solution to rejoin the solid. At the point where dissolving and "undissolving" are going on at the same rate, we arrive at a state of dynamic equilibrium. We write the equilibrium constant (the solubility product), K_{sp}, for the dissolving of a slightly soluble salt in the usual manner: since the concentration of the solid material is constant, we do not include it in the expression for K_{sp}:

$AgCl(s) \rightleftharpoons Ag^+(aq) + Cl^-(aq)$ $\qquad K_{sp} = [Ag^+][Cl^-]$

$PbCl_2(s) \rightleftharpoons Pb^{2+}(aq) + 2Cl^-(aq)$ $\qquad K_{sp} = [Pb^{2+}][Cl^-]^2$

$BaSO_4(s) \rightleftharpoons Ba^{2+}(aq) + SO_4^{2-}(aq)$ $\qquad K_{sp} = [Ba^{2+}][SO_4^{2-}]$

If the solubility product constant for a slightly soluble salt is known, the solubility of the salt in mol/L or g/L can be easily calculated. For $BaSO_4$, $K_{sp} = 1.5 \times 10^{-9}$ at 25°C. Suppose x moles of $BaSO_4$ dissolve per liter; since the stoichiometric coefficients for the dissolving of $BaSO_4$ are all the same, this means that x moles of $Ba^{2+}(aq)$ and x moles of $SO_4^{2-}(aq)$ will be produced per liter when $BaSO_4$ dissolves.

$K_{sp} = [Ba^{2+}][SO_4^{2-}] = [x][x] = 1.5 \times 10^{-9}$

$x^2 = 1.5 \times 10^{-9}$ and therefore $x = 3.9 \times 10^{-5}$ M

So the molar solubility of $BaSO_4$ is 3.9×10^{-5} M; this could be converted to the number of grams of $BaSO_4$ that dissolve per liter using the molar mass of $BaSO_4$ (233.4 g):

$(3.9 \times 10^{-5}$ mol/L$)(233.4$ g/mol$) = 9.10 \times 10^{-3}$ g/L

10. A conjugate acid-base pair consists of two species related to each other by the donating or accepting of a single proton, H^+. An acid has one more H^+ than its conjugate base; a base has one less H^+ than its conjugate acid.

Brönsted-Lowry acids:

$HCl(aq) + H_2O(l) \rightarrow Cl^-(aq) + H_3O^+(aq)$

$H_2SO_4(aq) + H_2O(l) \rightarrow HSO_4^-(aq) + H_3O^+(aq)$

$H_3PO_4(aq) + H_2O(l) \rightleftharpoons H_2PO_4^-(aq) + H_3O^+(aq)$

$NH_4^+(aq) + H_2O(l) \rightleftharpoons NH_3(aq) + H_3O^+(aq)$

Brönsted-Lowry bases:

$NH_3(aq) + H_2O(l) \rightleftharpoons NH_4^+(aq) + OH^-(aq)$

$HCO_3^-(aq) + H_2O(l) \rightleftharpoons H_2CO_3(aq) + OH^-(aq)$

$NH_2^-(aq) + H_2O(l) \rightarrow NH_3(aq) + OH^-(aq)$

$H_2PO_4^-(aq) + H_2O(l) \rightleftharpoons H_3PO_4(aq) + OH^-(aq)$

12. The strength of an acid is a direct result of the position of the acid's ionization equilibrium. We call acids whose ionization equilibrium position lies far to the right strong acids, and we call those acids whose equilibrium position lies only slightly to the right weak acids. For example, HCl, HNO$_3$, and HClO$_4$ are all strong acids, which means that they are completely ionized in aqueous solution (the position of equilibrium is very far to the right):

$$HCl(aq) + H_2O(l) \rightarrow Cl^-(aq) + H_3O^+(aq)$$
$$HNO_3(aq) + H_2O(l) \rightarrow NO_3^-(aq) + H_3O^+(aq)$$
$$HClO_4(aq) + H_2O(l) \rightarrow ClO_4^-(aq) + H_3O^+(aq)$$

Since these are very strong acids, we know that their anions (Cl$^-$, NO$_3^-$, ClO$_4^-$) must be very weak bases, and that solutions of the sodium salts of these anions would *not* be appreciably basic. Since these acids have a strong tendency to lose protons, there is very little tendency for the anions (bases) to gain protons.

14. The pH of a solution is defined as the negative of the base 10 logarithm of the hydrogen ion concentration in the solution; that is

$$pH = -\log_{10}[H^+]$$

for a solution. Since in pure water, the amount of H$^+$(aq) ion present is equal to the amount of OH$^-$(aq) ion, we say that pure water is *neutral*. Since [H$^+$] = 1.0 × 10^{-7} M in pure water, this means that the pH of pure water is -log[1.00 × 10^{-7} M] = 7.00. Solutions in which the hydrogen ion concentration is greater than 1.0 × 10^{-7} M (pH < 7.00) are *acidic*; solutions in which the hydrogen ion concentration is less than 1.00 × 10^{-7} M (pH > 7.00) are *basic*. Since the pH scale is logarithmic, when the pH changes by one unit, this corresponds to a change in the hydrogen ion concentration by a factor of *ten*.

In some instances, it may be more convenient to speak directly about the hydroxide ion concentration present in a solution, and so an analogous logarithmic expression is defined for the hydroxide ion concentration:

$$pOH = -\log_{10}[OH^-]$$

The concentrations of hydrogen ion and hydroxide ion in water (and in aqueous solutions) are *not* independent of one another, but rather are related by the dissociation equilibrium constant for water,

$$K_w = [H^+][OH^-] = 1.0 \times 10^{-14} \text{ at } 25°C.$$

From this constant it is obvious that pH + pOH = 14.00 for water (or an aqueous solution) at 25°C.

16. a. $K = [ClF]^2/[Cl_2][F_2]$ d. $K = 1/[NH_3][HCl]$
b. $K = [SO_3]^2/[SO_2]^2[O_2]$ e. $K = [H_2]/[C_2H_2]$
c. $K = [NO_2]^2/[N_2O_4]$

18. a. $K_{sp} = [Ca^{2+}][SO_4^{2-}]$
    b. $K_{sp} = [Pb^{2+}][Cl^-]^2$
    c. $K_{sp} = [Ca^{2+}]^3[PO_4^{3-}]^2$
    d. $K_{sp} = [Ni^{2+}][S^{2-}]$
    e. $K_{sp} = [Sn^{4+}][S^{2-}]^2$
    f. $K_{sp} = [Ba^{2+}][CrO_4^{2-}]$
    g. $K_{sp} = [Ag^+]^2[CrO_4^{2-}]$

20. a. $C_2H_3O_2^-$
    b. HS^-
    c. S^{2-}
    d. CO_3^{2-}
    e. HCO_3^-
    f. SO_4^{2-}
    g. NH_2^-
    h. HSO_4^-
    i. $H_2PO_4^-$

22. a. $[H^+] = 1.45 \times 10^{-6}$ M
       pH = $-\log[1.45 \times 10^{-6}]$ = 5.838
       pOH = 14 - pH = 14 - 5.838 = 8.161 = 8.16
    b. $[H^+] = 1.00 \times 10^{-14}/7.21 \times 10^{-3} = 1.39 \times 10^{-12}$ M = 1.4×10^{-12} M
    c. pOH = 14 - 5.981 = 8.019
       $[OH^-] = 10^{-8.019} = 9.57 \times 10^{-9}$ M = 9.6×10^{-9} M
    d. pH = 14 - 4.213 = 9.787
       $[H^+] = 10^{-9.787} = 1.63 \times 10^{-10}$ M = 1.6×10^{-10} M

Chapter 18   Oxidation-Reduction Reactions/Electrochemistry

2. Reduction can be defined as the gaining of electrons by an atom, molecule, or ion. Reduction may also be defined as a decrease in oxidation state for an element, but naturally such a decrease takes place by the gaining of electrons (so the two definitions are equivalent). The following equation shows the reduction of silver ion, which reaction is used in the jewelry business to plate silver metal on top of a cheaper metal.

$$Ag^+(aq) + e^- \rightarrow Ag(s)$$

Na → Na$^+$ + e$^-$ is an example of an oxidation process.

4. Each of these reactions involves a *metallic* element in the form of the *free* element on one side of the equation; on the other side of the equation, the metallic element is *combined* in an ionic compound. If a metallic element goes from the free metal to the ionic form, the metal is oxidized (loses electrons).

   a. sodium is oxidized, nitrogen is reduced
   b. magnesium is oxidized, chlorine is reduced
   c. aluminum is oxidized, bromine is reduced
   d. iron is oxidized, oxygen is reduced

6. Each of these reactions involves a *metallic* element in the form of the *free* element on one side of the equation; on the other side of the equation, the metallic element is *combined* in an ionic compound. If a metallic element goes from the free metal to the ionic form, the metal is oxidized (loses electrons).

   a. calcium is oxidized, sulfur is reduced
   b. copper is oxidized, fluorine is reduced
   c. calcium is oxidized, carbon is reduced
   d. zinc is oxidized, oxygen is reduced

8. The oxidation state of a *pure element* is *zero*, regardless of whether the element occurs naturally as single atoms or as a molecule.

10. Fluorine is always assigned a negative oxidation state (−1) because all other elements are less electronegative than fluorine. The other halogens are *usually* assigned an oxidation state of −1 in compounds. In interhalogen compounds such as ClF, fluorine is assigned oxidation state −1 (F is more electronegative than Cl), which means that chlorine must be assigned a +1 oxidation state in this instance.

12. Oxidation states represent a bookkeeping method to assign electrons in a molecule or ion. Since an ion has a net charge, the sum of the oxidation states of the atoms in the ion must equal the charge on the ion. For

230   Chapter 18   Oxidation-Reduction Reactions/Electrochemistry

example, the hydroxide ion (OH⁻) has an overall charge of −1 because hydrogen has an oxidation state of +1, whereas oxygen has an oxidation state of −2 in the hydroxide ion: (−2) + (+1) = −1

14. The rules for assigning oxidation states are given in Section 18.2 of the text. The rule which applies for each element in the following answers is given in parentheses after the element and its oxidation state.

   a.   N, +3 (Rule 6); Cl, −1 (Rule 5)
   b.   S, +6 (Rule 6); F, −1 (Rule 5)
   c.   P, +5 (Rule 6); Cl, −1 (Rule 5)
   d.   Si, −4 (Rule 6); H, +1 (Rule 4)

16. The rules for assigning oxidation states are given in Section 18.2 of the text. The rule which applies for each element in the following answers is given in parentheses after the element and its oxidation state.

   a.   Cl, 0 (Rule 1)
   b.   Cl, +1 (Rule 6); O, −2 (Rule 3)
   c.   H, +1 (Rule 4); Cl, −1 (Rule 5)
   d.   Cl, +1 (Rule 6); F, −1 (Rule 5)

18. The rules for assigning oxidation states are given in Section 18.2 of the text. The rule which applies for each element in the following answers is given in parentheses after the element and its oxidation state.

   a.   H, +1 (Rule 4); O, −2 (Rule 3); N, +5 (Rule 6)
   b.   H, +1 (Rule 4); O, −2 (Rule 3); P, +5 (Rule 7)
   c.   H, +1 (Rule 4); O, −2 (Rule 3); S. +6 (Rule 7)
   d.   O, −1 (Rule 3)

20. The rules for assigning oxidation states are given in Section 18.2 of the text. The rule which applies for each element in the following answers is given in parentheses after the element and its oxidation state.

   a.   Cu, +2 (Rules 2,6); Cl, −1 (Rules 2,5)
   b.   Cr, +3 (Rules 2,6); Cl, −1 (Rules 2,5)
   c.   H, +1 (Rule 4); O, −2 (Rule 3); Cr, +6 (Rule 7)
   d.   Cr, +3 (Rules 2,6); O, −2 (Rules 2,3)

22. The rules for assigning oxidation states are given in Section 18.2 of the text. The rule which applies for each element in the following answers is given in parentheses after the element and its oxidation state.

    a. Cl, +7 (Rule 7); O, −2 (Rule 3)

    b. Cl, +5 (Rule 7); O, −2 (Rule 3)

    c. Cl, +3 (Rule 7); O, −2 (Rule 3)

    d. Cl, +1 (Rule 7); O, −2 (Rule 3)

24. Electrons are negative; when an atom gains electrons, it gains one negative charge for each electron gained. For example, in the reduction reaction $Cl + e^- \rightarrow Cl^-$, the oxidation state of chlorine decreases from 0 to −1 as the electron is gained.

26. An oxidizing agent *causes* another species to be oxidized (to lose electrons). In order to make another species lose electrons, the oxidizing agent must be capable of gaining the electrons; an oxidizing agent is itself reduced. On the contrary, a reducing agent is itself oxidized.

28. An oxidizing agent oxidizes another species by gaining the electrons lost by the other species; therefore, an oxidizing agent itself decreases in oxidation state. A reducing agent increases its oxidation state when acting on another atom or molecule.

30. a. $2C(s) + O_2(g) \rightarrow 2CO(g)$

       C 0   O 0   C +2

                             O −2

      carbon is oxidized; oxygen is reduced

    b. $C_2H_2(g) + 2Br_2(l) \rightarrow C_2H_2Br_4(l)$

       C −1   Br 0   C +1

       H +1   H +1

                             Br −1

      carbon is oxidized; bromine is reduced

    c. $CH_4(g) + I_2(s) \rightarrow CH_3I(s) + HI(g)$

       C −4   I 0   C −2   H +1

       H +1   H +1   I −1

                      I −1

      carbon is oxidized; iodine is reduced

d. $P_4(s) + 6Cl_2(g) \rightarrow 4PCl_3(g)$
    P 0   Cl 0   P +3
                                Cl −3

phosphorus is oxidized; chlorine is reduced

32. a. $Cu(s) + 2AgNO_3(aq) \rightarrow 2Ag(s) + Cu(NO_3)_2(aq)$
    Cu 0   Ag +1   Ag 0   Cu +2
             N +5   N +5
             O −2   O −2

copper is oxidized; silver is reduced

b. $N_2(g) + 3F_2(g) \rightarrow 2NF_3(g)$
    N 0   F 0   F −1
                      N +3

nitrogen is oxidized; fluorine is reduced

c. $2Fe_2O_3(s) + 3S(s) \rightarrow 4Fe(s) + 3SO_2$
    Fe +3   S 0   Fe 0   S +4
    O −2   O −2

sulfur is oxidized; iron is reduced

d. $2H_2O_2(l) \rightarrow 2H_2O(l) + O_2(g)$
    H +1   H +1   O 0
    O −1   O −2

oxygen is both oxidized and reduced

34. Iron is reduced [+3 in $Fe_2O_3(s)$, 0 in $Fe(l)$]; carbon is oxidized [+2 in $CO(g)$, +4 in $CO_2(g)$]. $Fe_2O_3(s)$ is the oxidizing agent; $CO(g)$ is the reducing agent.

36. Chlorine is reduced [0 in $Cl_2(g)$, −1 in $NaCl(s)$]; Bromine is oxidized [−1 in $NaBr(aq)$, 0 in $Br_2(l)$]. $Cl_2(g)$ is the oxidizing agent; $NaBr(aq)$ is the reducing agent.

38. Oxidation-reduction reactions are often more complicated than "regular" reactions; frequently the coefficients necessary to balance the number of electrons transferred come out to be large numbers. We also have to make certain that we account for the electrons being transferred.

40. Under ordinary conditions it is impossible to have "free" electrons that are not part of some atom, ion, or molecule. For this reason, the total

number of electrons lost by the species being oxidized must equal the total number of electrons gained by the species being reduced.

42. a. $N_2(g) \rightarrow N^{3-}(s)$

   Balance nitrogen: $N_2(g) \rightarrow 2N^{3-}(s)$

   Balance charge: $6e^- + N_2(g) \rightarrow 2N^{3-}(s)$

   Balanced half reaction: $6e^- + N_2(g) \rightarrow 2N^{3-}(s)$

   b. $O_2^{2-}(aq) \rightarrow O_2(g)$

   Balance charge: $O_2^{2-}(aq) \rightarrow O_2(g) + 2e^-$

   Balanced half reaction: $O_2^{2-}(aq) \rightarrow O_2(g) + 2e^-$

   c. $Zn(s) \rightarrow Zn^{2+}(aq)$

   Balance charge: $Zn(s) \rightarrow Zn^{2+}(aq) + 2e^-$

   Balanced half reaction: $Zn(s) \rightarrow Zn^{2+}(aq) + 2e^-$

   d. $F_2(g) \rightarrow F^-(aq)$

   Balance fluorine: $F_2(g) \rightarrow 2F^-(aq)$

   Balance charge: $2e^- + F_2(g) \rightarrow 2F^-(aq)$

   Balanced half reaction: $2e^- + F_2(g) \rightarrow 2F^-(aq)$

44. a. $O_2(g) \rightarrow H_2O(l)$

   Balance oxygen: $O_2(g) \rightarrow 2H_2O(l)$

   Balance hydrogen: $4H^+(aq) + O_2(g) \rightarrow 2H_2O(l)$

   Balance charge: $4e^- + 4H^+(aq) + O_2(g) \rightarrow 2H_2O(l)$

   Balanced half reaction: $4e^- + 4H^+(aq) + O_2(g) \rightarrow 2H_2O(l)$

   b. $IO_3^-(aq) \rightarrow I_2(s)$

   Balance iodine: $2IO_3^-(aq) \rightarrow I_2(s)$

   Balance oxygen: $2IO_3^-(aq) \rightarrow I_2(s) + 6H_2O(l)$

   Balance hydrogen: $12H^+ + 2IO_3^-(aq) \rightarrow I_2(s) + 6H_2O(l)$

   Balance charge: $12H^+(aq) + 2IO_3^-(aq) + 10e^- \rightarrow I_2(s) + 6H_2O(l)$

   Balanced half reaction: $12H^+(aq) + 2IO_3^-(aq) + 10e^- \rightarrow I_2(s) + 6H_2O(l)$

   c. $VO^{2+}(aq) \rightarrow V^{3+}(aq)$

   Balance oxygen: $VO^{2+}(aq) \rightarrow V^{3+}(aq) + H_2O(l)$

   Balance hydrogen: $2H^+(aq) + VO^{2+}(aq) \rightarrow V^{3+}(aq) + H_2O(l)$

Balance charge: $e^- + 2H^+(aq) + VO^{2+}(aq) \rightarrow V^{3+}(aq) + H_2O(l)$

Balanced half reaction: $e^- + 2H^+(aq) + VO^{2+}(aq) \rightarrow V^{3+}(aq) + H_2O(l)$

d. $BiO^+(aq) \rightarrow Bi(s)$

Balance oxygen: $BiO^+(aq) \rightarrow Bi(s) + H_2O(l)$

Balance hydrogen: $2H^+(aq) + BiO^+(aq) \rightarrow Bi(s) + H_2O(l)$

Balance charge: $3e^- + 2H^+(aq) + BiO^+(aq) \rightarrow Bi(s) + H_2O(l)$

Balanced half reaction: $3e^- + 2H^+(aq) + BiO^+(aq) \rightarrow Bi(s) + H_2O(l)$

46. For simplicity, the physical states of the substances have been omitted until the final balanced equation is given.

a. $BrO_3^-(aq) + Br^-(aq) \rightarrow Br_2(l)$

$BrO_3^- \rightarrow Br_2$

Balance bromine: $2BrO_3^- \rightarrow Br_2$

Balance oxygen: $2BrO_3^- \rightarrow Br_2 + 6H_2O$

Balance hydrogen: $12H^+ + 2BrO_3^- \rightarrow Br_2 + 6H_2O$

Balance charge: $10e^- + 12H^+ + 2BrO_3^- \rightarrow Br_2 + 6H_2O$

Balanced half reaction: $10e^- + 12H^+ + 2BrO_3^- \rightarrow Br_2 + 6H_2O$

$Br^- \rightarrow Br_2$

Balance bromine: $2Br^- \rightarrow Br_2$

Balance charge: $2Br^- \rightarrow Br_2 + 2e^-$

Balanced half reaction: $2Br^- \rightarrow Br_2 + 2e^-$

Combine half reactions: $10e^- + 12H^+ + 2BrO_3^- \rightarrow Br_2 + 6H_2O$

$5 \times (2Br^- \rightarrow Br_2 + 2e^-)$

$12H^+(aq) + 2BrO_3^-(aq) + 10Br^-(aq) \rightarrow 6Br_2(l) + 6H_2O(l)$

$6H^+(aq) + BrO_3^-(aq) + 5Br^-(aq) \rightarrow 3Br_2(l) + 3H_2O(l)$

b. $I^-(aq) + O_2(g) \rightarrow I_2(s)$

$I^- \rightarrow I_2$

Balance iodine: $2I^- \rightarrow I_2$

Balance charge: $2I^- \rightarrow I_2 + 2e^-$

Balanced half reaction: $2I^- \rightarrow I_2 + 2e^-$

$O_2 \rightarrow H_2O$ (reaction takes place in acidic solution)

Balance oxygen: $O_2 \rightarrow 2H_2O$

Balance hydrogen: $4H^+ + O_2 \rightarrow 2H_2O$

Balance charge: $4e^- + 4H^+ + O_2 \rightarrow 2H_2O$

Balanced half reaction: $4e^- + 4H^+ + O_2 \rightarrow 2H_2O$

Combine half reactions: $4e^- + 4H^+ + O_2 \rightarrow 2H_2O$
$2 \times (2I^- \rightarrow I_2 + 2e^-)$

$4I^-(aq) + 4H^+(aq) + O_2(g) \rightarrow 2I_2(s) + 2H_2O(l)$

c. $Cr_2O_7^{2-}(aq) + I^-(aq) \rightarrow I_2(s) + Cr^{3+}(aq)$

$Cr_2O_7^{2-} \rightarrow Cr^{3+}$

Balance chromium: $Cr_2O_7^{2-} \rightarrow 2Cr^{3+}$

Balance oxygen: $Cr_2O_7^{2-} \rightarrow 2Cr^{3+} + 7H_2O$

Balance hydrogen: $14H^+ + Cr_2O_7^{2-} \rightarrow 2Cr^{3+} + 7H_2O$

Balance charge: $6e^- + 14H^+ + Cr_2O_7^{2-} \rightarrow 2Cr^{3+} + 7H_2O$

Balanced half reaction: $6e^- + 14H^+ + Cr_2O_7^{2-} \rightarrow 2Cr^{3+} + 7H_2O$

$I^- \rightarrow I_2$

Balance iodine: $2I^- \rightarrow I_2$

Balance charge: $2I^- \rightarrow I_2 + 2e^-$

Balanced half reaction: $2I^- \rightarrow I_2 + 2e^-$

Combine half reactions: $6e^- + 14H^+ + Cr_2O_7^{2-} \rightarrow 2Cr^{3+} + 7H_2O$
$3 \times (2I^- \rightarrow I_2 + 2e^-)$

$14H^+(aq) + Cr_2O_7^{2-}(aq) + 6I^-(aq) \rightarrow 2Cr^{3+}(aq) + 3I_2(s) + 7H_2O(l)$

d. $Mn^{3+}(aq) \rightarrow MnO_2(s) + Mn^{2+}(aq)$

$Mn^{3+} \rightarrow Mn^{2+}$

Balance charge: $e^- + Mn^{3+} \rightarrow Mn^{2+}$

Balanced half reaction: $e^- + Mn^{3+} \rightarrow Mn^{2+}$

$Mn^{3+} \rightarrow MnO_2$

Balance oxygen: $2H_2O + Mn^{3+} \rightarrow MnO_2$

Balance hydrogen: $2H_2O + Mn^{3+} \rightarrow MnO_2 + 4H^+$

Balance charge: $2H_2O + Mn^{3+} \rightarrow MnO_2 + 4H^+ + e^-$

Balanced half reaction: $2H_2O + Mn^{3+} \rightarrow MnO_2 + 4H^+ + e^-$

$2H_2O(l) + 2Mn^{3+}(aq) \rightarrow MnO_2(s) + Mn^{2+}(aq) + 4H^+(aq)$

48. The half reaction for the reduction of Ce^{4+} is the *same* in each of these processes:

$Ce^{4+} + e^- \rightarrow Ce^{3+}$

a. $H_3AsO_3 \rightarrow H_3AsO_4$

Balance oxygen: $H_2O + H_3AsO_3 \rightarrow H_3AsO_4$

Balance hydrogen: $H_2O + H_3AsO_3 \rightarrow H_3AsO_4 + 2H^+$

Balance charge: $H_2O + H_3AsO_3 \rightarrow H_3AsO_4 + 2H^+ + 2e^-$

Balanced half reaction: $H_2O + H_3AsO_3 \rightarrow H_3AsO_4 + 2H^+ + 2e^-$

Combine half reactions: $H_2O + H_3AsO_3 \rightarrow H_3AsO_4 + 2H^+ + 2e^-$
$$2 \times (Ce^{4+} + e^- \rightarrow Ce^{3+})$$

$H_2O(l) + H_3AsO_3(aq) + 2Ce^{4+} \rightarrow H_3AsO_4(aq) + 2H^+ + 2Ce^{3+}$

b. $Fe^{2+} \rightarrow Fe^{3+}$

Balance charge: $Fe^{2+} \rightarrow Fe^{3+} + e^-$

Balanced half reaction: $Fe^{2+} \rightarrow Fe^{3+} + e^-$

Combine half reactions: $Fe^{2+} \rightarrow Fe^{3+} + e^-$
$$Ce^{4+} + e^- \rightarrow Ce^{3+}$$

$Ce^{4+}(aq) + Fe^{2+}(aq) \rightarrow Ce^{3+}(aq) + Fe^{3+}(aq)$

c. $I^- \rightarrow I_2$

Balance iodine: $2I^- \rightarrow I_2$

Balance charge: $2I^- \rightarrow I_2 + 2e^-$

Balanced half reaction: $2I^- \rightarrow I_2 + 2e^-$

Combine half reactions: $2I^- \rightarrow I_2 + 2e^-$
$$2 \times (Ce^{4+} + e^- \rightarrow Ce^{3+})$$

$2Ce^{4+}(aq) + 2I^-(aq) \rightarrow 2Ce^{3+}(aq) + I_2(s)$

50. A salt bridge typically consists of a U-shaped tube filled with an inert electrolyte (one involving ions that are not part of the oxidation-reduction reaction). A salt bridge is used to complete the electrical circuit in a cell. Any method which allows transfer of charge without allowing bulk mixing of the solutions may be used.

52. In a galvanic cell, the anode is the electrode where oxidation occurs; the cathode is the electrode where reduction occurs.

54.

$Pb^{2+}(aq)$ ion is reduced; $Zn(s)$ is oxidized.

The reaction at the anode is $Zn(s) \rightarrow Zn^{2+}(aq) + 2e^-$.

The reaction at the cathode is $Pb^{2+}(aq) + 2e^- \rightarrow Pb(s$

56. Both normal and alkaline cells contain zinc as one electrode; zinc corrodes more slowly under alkaline conditions than in the highly acidic environment of a normal dry cell.

anode: $Zn(s) + 2OH^-(aq) \rightarrow ZnO(s) + H_2O(l) + 2e^-$

cathode: $2MnO_2(s) + H_2O(l) + 2e^- \rightarrow Mn_2O_3(s) + 2OH^-(aq)$

58. Aluminum is a very reactive metal when freshly isolated in the pure state. However, on standing for even a relatively short period of time, aluminum metal forms a thin coating of Al_2O_3 on its surface from reaction with atmospheric oxygen. This coating of Al_2O_3 is much less reactive than the metal and serves to protect the surface of the metal from further attack.

60. In cathodic protection of steel tanks and pipes, a more reactive metal than iron is connected to the item to be protected. The active metal is then preferentially oxidized rather than the iron of the tank or pipe.

62. The main recharging reaction for the lead storage battery is

$$2PbSO_4(s) + 2H_2O(l) \rightarrow Pb(s) + PbO_2(s) + 2H_2SO_4(aq)$$

A major side reaction is the electrolysis of water

$$2H_2O(l) \rightarrow 2H_2(g) + O_2(g)$$

resulting in the production of an explosive mixture of hydrogen and oxygen, which accounts for many accidents during the recharging of such batteries.

64. Electrolysis is applied in electroplating by making the item to be plated the cathode in a cell containing a solution of ions of the desired plating metal.

66. loss; oxidation state

68. electronegative

70. zero

72. lose

74. separate from

76. anode

78. electrolysis

80. hydrogen; oxygen

82. oxidation

84. a. $4Fe(s) + 3O_2(g) \rightarrow 2Fe_2O_3(s)$
iron is oxidized, oxygen is reduced
b. $2Al(s) + 3Cl_2(g) \rightarrow 2AlCl_3(s)$
aluminum is oxidized, chlorine is reduced
c. $6Mg(s) + P_4(s) \rightarrow 2Mg_3P_2(s)$
magnesium is oxidized, phosphorus is reduced

86. a. magnesium is oxidized, aluminum (as Al^{3+}) is reduced
b. magnesium is oxidized, aluminum (as Al^{3+} in $AlCl_3$) is reduced
c. oxygen is oxidized, chlorine is reduced

88. a. $C_3H_8(g) + 5O_2(g) \rightarrow 3CO_2(g) + 4H_2O(g)$
b. $CO(g) + 2H_2(g) \rightarrow CH_3OH(l)$
c. $SnO_2(s) + 2C(s) \rightarrow Sn(s) + 2CO(g)$
d. $C_2H_5OH(l) + 3O_2(g) \rightarrow 2CO_2(g) + 3H_2O(g)$

90. Each of these reactions involves a *metallic* element in the form of the *free* element on one side of the equation; on the other side of the equation, the metallic element is *combined* in an ionic compound. If a metallic element goes from the free metal to the ionic form, the metal is oxidized (loses electrons).

a. sodium is oxidized, oxygen is reduced
b. iron is oxidized, hydrogen is reduced
c. oxygen (O^{2-}) is oxidized, aluminum (Al^{3+}) is reduced (this reaction is the reverse of the type discussed above)
d. magnesium is oxidized, nitrogen is reduced

92. The rules for assigning oxidation states are given in Section 18.2 of the text. The rule which applies for each element in the following answers is given in parentheses after the element and its oxidation state.

 a. H +1 (Rule 4); N −3 (Rule 6)
 b. C +2 (Rule 6); O −2 (Rule 3)
 c. C +4 (Rule 6); O −2 (Rule 3)
 d. N +3 (Rule 6); F −1 (Rule 5)

94. The rules for assigning oxidation states are given in Section 18.2 of the text. The rule which applies for each element in the following answers is given in parentheses after the element and its oxidation state.

 a. S +4 (Rule 6); O −2 (Rule 3)
 b. S +6 (Rule 6); O −2 (Rule 3)
 c. H +1 (Rule 4); S −2 (Rule 6)
 d. H +1 (Rule 4); S +6 (Rule 6); O −2 (Rule 3)

96. The rules for assigning oxidation states are given in Section 18.2 of the text. The rule which applies for each element in the following answers is given in parentheses after the element and its oxidation state.

 a. Bi +3 (Rule 7); O −2 (Rule 3)
 b. P +5 (Rule 7); O −2 (Rule 3)
 c. N +3 (Rule 7); O −2 (Rule 3)
 d. Hg +1 (Rule 7)

98. a. $2B_2O_3(s) + 6Cl_2(g) \rightarrow 4BCl_3(l) + 3O_2(g)$
    B +3   Cl 0   B +3   O 0
    O −2   Cl −1

    oxygen is oxidized; chlorine is reduced

 b. $GeH_4(g) + O_2(g) \rightarrow Ge(s) + 2H_2O(g)$
    Ge −4   O 0   Ge 0   H +1
    H +1   O −2

    germanium is oxidized; oxygen is reduced

c. $C_2H_4(g) + Cl_2(g) \rightarrow C_2H_4Cl_2(l)$

    C −2   Cl 0   C −1

    H +1   H +1

                                Cl −1

carbon is oxidized; chlorine is reduced

d. $O_2(g) + 2F_2(g) \rightarrow 2OF_2(g)$

    O 0   F 0   O +2

                              F −1

oxygen is oxidized; fluorine is reduced

100. a. $SiO_2(s) \rightarrow Si(s)$

Balance oxygen: $SiO_2(s) \rightarrow Si(s) + 2H_2O(l)$

Balance hydrogen: $SiO_2(s) + 4H^+(aq) \rightarrow Si(s) + 2H_2O(l)$

Balance charge: $SiO_2(s) + 4H^+(aq) + 4e^- \rightarrow Si(s) + 2H_2O(l)$

Balanced half reaction: $SiO_2(s) + 4H^+(aq) + 4e^- \rightarrow Si(s) + 2H_2O(l)$

b. $S(s) \rightarrow H_2S(g)$

Balance hydrogen: $S(s) + 2H^+(aq) \rightarrow H_2S(g)$

Balance charge: $S(s) + 2H^+(aq) + 2e^- \rightarrow H_2S(g)$

Balanced half reaction: $S(s) + 2H^+(aq) + 2e^- \rightarrow H_2S(g)$

c. $NO_3^-(aq) \rightarrow HNO_2(aq)$

Balance oxygen: $NO_3^-(aq) \rightarrow HNO_2(aq) + H_2O(l)$

Balance hydrogen: $NO_3^-(aq) + 3H^+(aq) \rightarrow HNO_2(aq) + H_2O(l)$

Balance charge: $NO_3^-(aq) + 3H^+(aq) + 2e^- \rightarrow HNO_2(aq) + H_2O(l)$

Balanced half reaction: $NO_3^-(aq) + 3H^+(aq) + 2e^- \rightarrow HNO_2(aq) + H_2O(l)$

d. $NO_3^-(aq) \rightarrow NO(g)$

Balance oxygen: $NO_3^-(aq) \rightarrow NO(g) + 2H_2O(l)$

Balance hydrogen: $NO_3^-(aq) + 4H^+(aq) \rightarrow NO(g) + 2H_2O(l)$

Balance charge: $NO_3^-(aq) + 4H^+(aq) + 3e^- \rightarrow NO(g) + 2H_2O(l)$

Balanced half reaction: $NO_3^-(aq) + 4H^+(aq) + 3e^- \rightarrow NO(g) + 2H_2O(l)$

Chapter 18 Oxidation-Reduction Reactions/Electrochemistry 241

102. For simplicity, the physical states of the substances have been omitted until the final balanced equation is given.

For the reduction of the permanganate ion, MnO_4^-, in acid solution, the half reaction is always the *same*:

$MnO_4^- \rightarrow Mn^{2+}$

Balance oxygen: $MnO_4^- \rightarrow Mn^{2+} + 4H_2O$

Balance hydrogen: $8H^+ + MnO_4^- \rightarrow Mn^{2+} + 4H_2O$

Balance charge: $8H^+ + MnO_4^- + 5e^- \rightarrow Mn^{2+} + 4H_2O$

a. $C_2O_4^{2-} \rightarrow CO_2$

Balance carbon: $C_2O_4^{2-} \rightarrow 2CO_2$

Balance charge: $C_2O_4^{2-} \rightarrow 2CO_2 + 2e^-$

Combine half reactions: $5 \times (C_2O_4^{2-} \rightarrow 2CO_2 + 2e^-)$

$\qquad\qquad\qquad\qquad\qquad 2 \times (8H^+ + MnO_4^- + 5e^- \rightarrow Mn^{2+} + 4H_2O)$

$16H^+(aq) + 2MnO_4^-(aq) + 5C_2O_4^{2-}(aq) \rightarrow 2Mn^{2+}(aq) + 8H_2O(l) + 10CO_2(g)$

b. $Fe^{2+} \rightarrow Fe^{3+}$

Balance charge: $Fe^{2+} \rightarrow Fe^{3+} + e^-$

Combine half reactions: $5 \times (Fe^{2+} \rightarrow Fe^{3+} + e^-)$

$\qquad\qquad\qquad\qquad\qquad 8H^+ + MnO_4^- + 5e^- \rightarrow Mn^{2+} + 4H_2O$

$8H^+(aq) + MnO_4^-(aq) + 5Fe^{2+}(aq) \rightarrow Mn^{2+}(aq) + 4H_2O(l) + 5Fe^{3+}(aq)$

c. $Cl^- \rightarrow Cl_2$

Balance chlorine: $2Cl^- \rightarrow Cl_2$

Balance charge: $2Cl^- \rightarrow Cl_2 + 2e^-$

Combine half reactions: $5 \times (2Cl^- \rightarrow Cl_2 + 2e^-)$

$\qquad\qquad\qquad\qquad\qquad 2 \times (8H^+ + MnO_4^- + 5e^- \rightarrow Mn^{2+} + 4H_2O)$

$16H^+(aq) + 2MnO_4^-(aq) + 10Cl^-(aq) \rightarrow 2Mn^{2+}(aq) + 8H_2O(l) + 5Cl_2(g)$

Chapter 19   Radioactivity and Nuclear Energy

2. The radius of a typical atomic nucleus is on the order of 10^{-13} cm, which is about one hundred thousand times smaller than the radius of an atom overall.

4. The atomic number (Z) of a nucleus represents the number of positively-charged protons present in the nucleus. The mass number (A) of a nucleus represents the total number of protons and neutrons in the nucleus. For example, for the nuclide $^{13}_{6}C$, with six protons and seven neutrons, we have $Z = 6$ and $A = 13$.

6. The atomic number (Z) is written in such formulas as a left subscript, while the mass number (A) is written as a left superscript. That is, the general symbol for a nuclide is $^{A}_{Z}X$. As an example, consider the isotope of oxygen with 8 protons and 8 neutrons: its symbol would be $^{16}_{8}O$.

8. Alpha particle: charge 2+, mass number 4, symbol $^{4}_{2}He$

10. When a nucleus produces a beta (β) particle, the atomic number of the parent nucleus is *increased* by *one* unit. The beta particle has a mass number of zero, but an "atomic number" of −1.

12. Gamma rays are high energy photons of electromagnetic radiation. Gamma rays are not normally considered to be particles. When a nucleus produces only gamma radiation, the atomic number and mass number of the nucleus do not change. Gamma rays represent the energy changes associated with transitions and rearrangement of the particles within the nucleus.

14. Electron capture occurs when one of the inner orbital electrons is pulled into and becomes part of the nucleus.

16. The average atomic mass of lithium is 6.941, which is considerably closer to 7 than to 6, so the isotope of mass number 7 must predominate. The average atomic mass given in the periodic table is a *weighted* average which includes not only the masses of the isotopes, but also their relative abundances.

18. $^{32}_{16}S$ (16 protons, 16 neutrons)

    $^{33}_{16}S$ (16 protons, 17 neutrons)

    $^{34}_{16}S$ (16 protons, 18 neutrons)

    $^{36}_{16}S$ (16 protons, 20 neutrons)

20. a. $^{0}_{-1}e$ or $^{0}_{-1}\beta$

b. $^{0}_{+1}e$ or $^{0}_{+1}\beta$

c. $^{0}_{0}\gamma$

22. a. $^{0}_{+1}e$ (positron)

b. $^{226}_{88}Ra$

c. $^{234}_{90}Th$

24. a. $^{0}_{-1}e$

b. $^{0}_{-1}e$

c. $^{210}_{83}Bi$

26. a. $^{59}_{26}Fe \rightarrow ^{59}_{27}Co + ^{0}_{-1}e$

b. $^{24}_{11}Na \rightarrow ^{24}_{12}Mg + ^{0}_{-1}e$

c. $^{198}_{79}Au \rightarrow ^{198}_{80}Hg + ^{0}_{-1}e$

d. $^{47}_{20}Ca \rightarrow ^{47}_{21}Sc + ^{0}_{-1}e$

28. a. $^{226}_{88}Ra \rightarrow ^{222}_{86}Rn + ^{4}_{2}He$

b. $^{222}_{86}Rn \rightarrow ^{218}_{84}Po + ^{4}_{2}He$

c. $^{239}_{94}Pu \rightarrow ^{235}_{92}U + ^{4}_{2}He$

d. $^{8}_{4}Be \rightarrow ^{4}_{2}He + ^{4}_{2}He$

30. There is often considerable repulsion between the target nucleus and the particles being used for bombardment (especially if the bombarding

32. $^{27}_{13}\text{Al} + ^{4}_{2}\text{He} \rightarrow ^{30}_{15}\text{P} + ^{1}_{0}\text{n}$

34. The half-life of a nucleus is the time required for one-half of the original sample of nuclei to decay. A given isotope of an element always has the same half-life, although different isotopes of the same element may have greatly different half-lives. Nuclei of different elements typically have different half-lives.

36. $^{226}_{88}\text{Ra}$ is the most stable (longest half-life)

    $^{224}_{88}\text{Ra}$ is the "hottest" (shortest half-life)

38. highest activity $\qquad\qquad$ lowest activity

    $^{87}\text{Sr} > ^{99}\text{Tc} > ^{24}\text{Na} > ^{99}\text{Mo} > ^{133}\text{Xe} > ^{131}\text{I} > ^{32}\text{P} > ^{51}\text{Cr} > ^{59}\text{Fe}$

40. For ^{223}Ra, the half-life is 12 days. After two half-lives (24 days), 250 mg remains; after three half-lives (36 days), 125 mg remains.

    For ^{224}Ra, the half-life is 3.6 days. One month would be approximately 8 half-life periods (29 days), and approximately 4 mg remains.

    For ^{225}Ra, the half-life is 15 days. One month would be two half-life periods, and 250 mg remains.

42. For an administered dose of 100 μg, 0.39 μg remains after 2 days. The fraction remaining is 0.39/100 = 0.0039; on a percentage basis, less than 0.4 % of the original radioisotope remains.

44. Carbon-14 is produced in the upper atmosphere by the bombardment of ordinary nitrogen with neutrons from space:

    $^{14}_{7}\text{N} + ^{1}_{0}\text{n} \rightarrow ^{14}_{6}\text{C} + ^{1}_{1}\text{H}$

46. We assume that the concentration of C-14 in the atmosphere is effectively constant. A living organism is constantly replenishing C-14 either through the processes of metabolism (sugars ingested in foods contain C-14), or photosynthesis (carbon dioxide contains C-14). When a plant dies, it no longer replenishes itself with C-14 from the atmosphere, and as the C-14 undergoes radioactive decay, its amount decreases with time.

48. These isotopes and their uses are listed in Table 19.4 Some important examples include the use of I-131 (and other iodine isotopes) in the

diagnosis and treatment of thyroid disease (iodine is used in the body primarily in the thyroid gland); Fe-59 in the study of the function of red blood cells (iron is a constituent of hemoglobin which is found in the red blood cells); Sr-87 in the study of bones (Sr is a Group 2 element, and is able to take the place of Ca in bone structures).

50. Combining two light nuclei to form a heavier, more stable nucleus is called nuclear *fusion*. Splitting a heavy nucleus into nuclei with smaller mass numbers is called nuclear *fission*.

52. $^{1}_{0}n + ^{235}_{92}U \rightarrow ^{142}_{56}Ba + ^{91}_{36}Kr + 3\,^{1}_{0}n$ is one possibility.

54. A critical mass of a fissionable material is the amount needed to provide a high enough internal neutron flux to sustain the chain reaction (enough neutrons are produced to cause the continuous fission of further material). A sample with less than a critical mass is still radioactive, but cannot sustain a chain reaction.

56. An actual nuclear explosion, of the type produced by a nuclear weapon, cannot occur in a nuclear reactor because the concentration of the fissionable materials is not sufficient to form a supercritical mass. However, since many reactors are cooled by water, which can decompose into hydrogen and oxygen gases, a *chemical* explosion is possible which could scatter the radioactive material used in the reactor.

58. $^{238}_{92}U + ^{1}_{0}n \rightarrow ^{239}_{92}U$

   $^{239}_{92}U \rightarrow ^{239}_{93}Np + ^{0}_{-1}e$

   $^{239}_{93}Np \rightarrow ^{239}_{94}Pu + ^{0}_{-1}e$

60. In one type of fusion reactor, two $^{2}_{1}H$ atoms are fused to produce $^{4}_{2}He$. Because the hydrogen nuclei are positively charged, extremely high energies (temperatures of 40 million K) are needed to overcome the repulsion between the nuclei as they are shot into each other.

62. In the theory of stellar nucleosynthesis, it is considered that the nucleus began as a cloud of neutrons which exploded (the Big Bang). After this initial explosion, neutrons were thought to have decomposed into protons and electrons

   $^{1}_{0}n \rightarrow ^{1}_{1}H + ^{0}_{-1}e$

The products of this decomposition were then thought to have combined to form large clouds of hydrogen atoms. As the hydrogen clouds became larger, gravitational forces caused these clouds to contract and heat up. Eventually the clouds of hydrogen were so dense and so hot that fusion of hydrogen nuclei into helium nuclei took place, with a great

246   Chapter 19   Radioactivity and Nuclear Energy

release of energy. When the tendency for the hydrogen clouds to expand from the heat of fusion was counter-balanced by the gravitational forces of the cloud, a small star had formed. In addition to the fusion of hydrogen nuclei into helium mentioned already, as the star's hydrogen supply is exhausted, the helium present in the star also begins to undergo fusion into nuclei of other elements.

64. Somatic damage is damage directly to the organism itself, causing nearly immediate sickness or death to the organism. Genetic damage is damage to the genetic machinery of the organism, which will be manifested in future generations of offspring.

66. Gamma rays penetrate long distances, but seldom cause ionization of biological molecules. Alpha particles, because they are much heavier although less penetrating, are very effective at ionizing biological molecules and leave a dense trail of damage in the organism. Isotopes which release alpha particles can be ingested or breathed into the body where the damage from the alpha particles will be more acute.

68. The exposure limits given in Table 19.5 as causing no detectable clinical effect are 0–25 rem. The total yearly exposures from natural and human-induced radioactive sources are estimated in Table 19.6 as less than 200 *milli*rem (0.2 rem), which is well within the acceptable limits.

70. radioactive

72. mass

74. neutron; proton

76. radioactive decay

78. mass number

80. transuranium

82. half-life

84. radiotracers

86. chain

88. breeder

90. 4.5×10^9 dollars ($4.5 billion)

92. 3.5×10^{-11} J/atom; 8.9×10^{10} J/g

94. Despite the fact that nuclear waste has been generated for over 40 years, no permanent disposal plan has been implemented as yet. One proposal to dispose of such waste calls for the waste to be sealed in

Chapter 19   Radioactivity and Nuclear Energy   247

blocks of glass, which in turn are sealed in corrosion-proof metal drums, which would then be buried in deep, stable rock formations away from earthquake and other geologically active zones. In these deep storage areas, it is hoped that the waste could decay safely undisturbed until the radioactivity drops to "safe" levels.

96. $^{27}_{13}$Al: 13 protons, 14 neutrons

   $^{28}_{13}$Al: 13 protons, 15 neutrons

   $^{29}_{13}$Al: 13 protons, 16 neutrons

98. a. $^{1}_{0}n$

   b. $^{201}_{79}Au$

   c. $^{0}_{-1}e$

100. For a decay of 10 μg to 1/1000 of this amount, we want to know when the amount of remaining ^{131}I is on the order of 0.01 μg.

time, days	0	8	16	24	32	40	48	56	64	72	80
mass, μg	10	5	2.5	1.25	0.625	0.313	0.156	0.078	0.039	0.020	0.01

Approximately 80 days are required.

102. Breeder reactors are set up to convert non-fissionable ^{238}U into fissionable ^{239}Pu. The material used for fission in a breeder reactor is a combination of U-235 (which undergoes fission in a chain reaction) and the more common U-238 isotope. Excess neutrons from the U-235 fission are absorbed by the U-238 converting it to the fissionable plutonium isotope Pu-239. Although Pu-239 is fissionable, its chemical and physical properties make it very difficult and expensive to handle and process.

Chapter 20 Organic Chemistry

2. A given carbon atom can be attached to a maximum of *four* other atoms. Carbon atoms have four valence electrons. By making four bonds, carbon atoms exactly complete their valence octet.

4. A triple bond represents the sharing of *six* electrons (three *pairs* of electrons). The simplest example of an organic molecule containing a triple bond is acetylene, H:C:::C:H.

6. Each carbon atom in these hydrocarbons is bonded to *four* other atoms. According to VSEPR theory, each carbon atom has its electron pairs arranged *tetrahedrally*.

8. An unsaturated hydrocarbon is one that contains an area of multiple bonding between some of the carbon atoms. Such a hydrocarbon is called "unsaturated" because each carbon atom of the multiple bond could bond to additional atoms (rather than to each other). Such unsaturated hydrocarbons undergo addition reactions in which atoms or groups from an outside reagent become bonded to the carbon atoms of the multiple bond, producing a saturated compound. The structural features that characterize unsaturated hydrocarbons are the carbon-carbon double bond or triple bond.

10. Each successive member of this family differs from the previous member by a –CH$_2$– unit (which is sometimes called a "methylene" unit). Such a family of compounds is sometimes referred to as a *homologous series*.

```
      H                 H H              H H H             H H H H
      |                 | |              | | |             | | | |
   H–C–H            H–C–C–H          H–C–C–C–H         H–C–C–C–C–H
      |                 | |              | | |             | | | |
      H                 H H              H H H             H H H H
   methane            ethane            propane             butane


   H H H H H         H H H H H H         H H H H H H H
   | | | | |         | | | | | |         | | | | | | |
 H–C–C–C–C–C–H     H–C–C–C–C–C–C–H     H–C–C–C–C–C–C–C–H
   | | | | |         | | | | | |         | | | | | | |
   H H H H H         H H H H H H         H H H H H H H
     pentane            hexane              heptane


 H H H H H H H H      H H H H H H H H H      H H H H H H H H H H
 | | | | | | | |      | | | | | | | | |      | | | | | | | | | |
H–C–C–C–C–C–C–C–C–H  H–C–C–C–C–C–C–C–C–C–H  H–C–C–C–C–C–C–C–C–C–C–H
 | | | | | | | |      | | | | | | | | |      | | | | | | | | | |
 H H H H H H H H      H H H H H H H H H      H H H H H H H H H H
      octane                nonane                   decane
```

12. a.
```
        H H H H H H H H
        | | | | | | | |
      H–C–C–C–C–C–C–C–C–H
        | | | | | | | |
        H H H H H H H H
```

b.
```
    H H H H H H H H H H
    | | | | | | | | | |
  H-C-C-C-C-C-C-C-C-C-C-H
    | | | | | | | | | |
    H H H H H H H H H H
```

c.
```
    H H H H H H H H H H H H
    | | | | | | | | | | | |
  H-C-C-C-C-C-C-C-C-C-C-C-C-H
    | | | | | | | | | | | |
    H H H H H H H H H H H H
```

14. a.
```
    H H H H
    | | | |
  H-C-C-C-C-H
    | | | |
    H H H H
```
butane $CH_3CH_2CH_2CH_3$

b.
```
    H H H H H H H H
    | | | | | | | |
  H-C-C-C-C-C-C-C-C-H
    | | | | | | | |
    H H H H H H H H
```
octane $CH_3CH_2CH_2CH_2CH_2CH_2CH_2CH_3$

c.
```
    H H H H H
    | | | | |
  H-C-C-C-C-C-H
    | | | | |
    H H H H H
```
pentane $CH_3CH_2CH_2CH_2CH_3$

d.
```
    H H H H H H
    | | | | | |
  H-C-C-C-C-C-C-H
    | | | | | |
    H H H H H H
```
hexane $CH_3CH_2CH_2CH_2CH_2CH_3$

16. branch or substituent

18. With eight carbon atoms, there are a *lot* of isomers possible. Here are carbon skeletons for some of them:

```
C-C-C-C-C-C-C-C     C-C-C-C-C-C-C     C-C-C-C-C-C-C     C-C-C-C-C-C-C
                          |                   |                   |
                          C                   C                   C
```

```
        C                                                              C
        |                                                              |
    C-C-C-C-C-C      C-C-C-C-C-C      C-C-C-C-C-C      C-C-C-C-C
        |                |   |            |   |            |   |
        C                C   C            C   C            C   C


        C                C                C   C
        |                |                |   |
    C-C-C-C-C        C-C-C-C-C        C-C-C-C       (plus 7 others)
    |   |                |   |            |   |
    C   C                C   C            C   C
```

20. The root name is derived from the number of carbon atoms in the *longest continuous chain* of carbon atoms.

22. The position of a substituent is indicated by a number which corresponds to the carbon atom in the longest chain to which the substituent is attached.

24. Multiple substituents are listed in alphabetical order, disregarding any prefix.

26. a. 3-ethylpentane

    b. 2,2-dimethylbutane

    c. 2,2-dimethylpropane

    d. 2,3,4-trimethylpentane

28. a.
```
    CH₃-CH₂-CH-CH₂-CH₃
            |
           CH₃
```

    b.
```
    CH₃-CH₂-CH₂-CH-CH₂-CH₂-CH₃
                |
            CH₃-C-CH₃
                |
               CH₃
```

    c.
```
    CH₃-CH₂-CH₂-CH₂-CH-CH₂-CH₂-CH₂-CH₃
                    |
                CH₂-CH₂-CH₃
```

    d.
```
    CH₃-CH-CH₂-CH-CH₂-CH-CH₂-CH₂-CH-CH₃
        |       |       |           |
       CH₃     CH₃     CH₃         CH₃
```

30.
Number of C atoms	Use
C₅–C₁₂	gasoline
C₁₀–C₁₈	kerosene, jet fuel
C₁₅–C₂₅	diesel fuel, heating oil, lubrication
C₂₅–	asphalt

32. Tetraethyl lead was added to gasolines to prevent "knocking" of high efficiency automobile engines. The use of tetraethyl lead is being phased out because of the danger to the environment of the lead in this substance.

34. Combustion represents the vigorous reaction of a hydrocarbon (or other substance) with oxygen. The combustion of alkanes has been made use of as a source of heat, light, and mechanical energy.

36. Dehydrogenation reactions involve the removal of hydrogen atoms from adjacent carbon atoms in an alkane (or other substance). When two hydrogen atoms are removed from an alkane or related compound, a double bond is created.

38. a. $2C_2H_6(g) + 7O_2(g) \rightarrow 4CO_2(g) + 6H_2O(g)$

   b. $CH_2Cl_2(l) + Cl_2(g) \rightarrow HCl(g) + CHCl_3(l)$

   c. $CH_2=CH-CH_2-CH_3(l) + Br_2(l) \rightarrow CH_2-CH-CH_2-CH_3(l)$
   $\qquad\qquad\qquad\qquad\qquad\qquad\qquad\qquad\quad\;\;|\quad\;\;|$
   $\qquad\qquad\qquad\qquad\qquad\qquad\qquad\qquad\;\;Br\;\;Br$

40. An alkyne is a hydrocarbon containing a carbon-carbon triple bond. The general formula is C_nH_{2n-2}.

42. The location of a double or triple bond in the longest chain of an alkene or alkyne is indicated by giving the *number* of the lowest number carbon atom involved in the double or triple bond.

44. hydrogenation

46. a. propyne

   b. 3-hexyne

   c. 3-hexene

48. Shown are carbon skeletons:

   C≡C-C-C-C-C   C-C≡C-C-C-C   C-C-C≡C-C-C   C≡C-C-C-C
   $\qquad\qquad\qquad\qquad\qquad\qquad\qquad\qquad\qquad\qquad\;\;|\qquad\qquad\qquad\qquad\;|$
   $\qquad\qquad\qquad\qquad\qquad\qquad\qquad\qquad\qquad\;\;C\qquad\qquad\qquad\qquad\;C$
   $\qquad\qquad\qquad\qquad\qquad\qquad\qquad\qquad\qquad\;\;|$

   C-C≡C-C-C   C≡C-C-C-C   C≡C-C-C
   $\;\;\;\;\;|\qquad\qquad\qquad\;\;|\qquad\qquad\qquad\;|$
   $\;\;\;\;\;C\qquad\qquad\qquad\;C\qquad\qquad\qquad C$

50. For benzene, a *set* of equivalent Lewis structures can be drawn, differing only in the *location* of the three double bonds in the ring. Experimentally, however, benzene does not demonstrate the chemical properties expected for molecules having *any* double bonds. We say that the "extra" electrons that would go into making the second bond of the

three double bonds are delocalized around the entire benzene ring: this delocalization of the electrons explains benzene's unique properties.

52. When named as a substituent, the benzene ring is called the *phenyl* group. Two examples are

        CH$_2$=CH—CH—CH$_3$ CH$_3$—CH—CH$_2$—CH$_2$—CH$_2$—CH$_3$
                (phenyl)   (phenyl)

        3-phenyl-1-butene   2-phenylhexane

54. *ortho-* refers to adjacent substituents (1,2-); *meta-* refers to two substituents with one unsubstituted carbon atom between them (1,3-); *para-* refers to two substituents with two unsubstituted carbon atoms between them (1,4-).

56. a. 1,2-dimethylbenzene (*o*-xylene is its common name)
    b. 1,2,3,4,5,6-hexachlorobenzene
    c. anthracene
    d. 3,5-dichloro-1-methylbenzene (3,5-dichlorotoluene is another name)

58. a. ether
    b. alcohol
    c. alcohol
    d. organic (carboxylic) acid

60. Primary alcohols have *one* hydrocarbon fragment (alkyl group) bonded to the carbon atom where the -OH group is attached. Secondary alcohols have *two* such alkyl groups attached, and tertiary alcohols contain *three* such alkyl groups. Examples are

    ethanol (primary)   CH$_3$–CH$_2$–OH

    2-propanol (secondary)   CH$_3$–CH(OH)–CH$_3$

    2-methyl-2-propanol (tertiary)   (CH$_3$)$_3$C–OH

Chapter 20   Organic Chemistry   253

62. a. $CH_3-CH_2-CH_2-OH$ primary

   b. $CH_2-CH-CH_2-CH_2-CH_3$ secondary
         $|$
         OH

   c. $CH_3-CH-CH_2-OH$ primary
         $|$
         CH_3

   d. OH
         $|$
       $CH_2-CH_2-C-CH_2-CH_3$ tertiary
         $|$
         CH_3

64. $C_6H_{12}O_6$ —yeast→ $2CH_3-CH_2-OH + 2CO_2$

   The yeast necessary for the fermentation process are killed if the concentration of ethanol is over 13%. More concentrated ethanol solutions are most commonly made by distillation.

66. methanol (CH_3OH) - starting material for synthesis of acetic acid and many plastics

   ethylene glycol (CH_2OH-CH_2OH) - automobile antifreeze

   isopropyl alcohol (2-propanol, $CH_3-CH(OH)-CH_3$) - rubbing alcohol

68. Aldehydes and ketones both contain the carbonyl group (C=O).

   Aldehydes and ketones differ in the *location* of the carbonyl function: aldehydes contain the carbonyl group at the end of a hydrocarbon chain (the carbon atom of the carbonyl group is bonded only to at most one other carbon atom); the carbonyl group of ketones represents one of the interior carbon atoms of a chain (the carbon atom of the carbonyl group is bonded to two other carbon atoms).

70. Aldehydes and ketones are produced by the oxidation of primary and secondary alcohols, respectively.

   CH_3-CH_2-OH ---oxidation--→ $CH_3-C=O$
   $|$
   H

   $CH_3-CH-OH$ ---oxidation--→ $CH_3-C=O$
   $||$
   CH_3CH_3

72. In addition to their systematic names (based on the hydrocarbon root, with the ending *-one*), ketones can also be named by naming the groups

attached to either side of the carbonyl carbon as alkyl groups, followed by the word "ketone". Examples are

CH$_3$–C(=O)–CH$_2$CH$_3$ methyl ethyl ketone (2-butanone, butanone)

CH$_3$CH$_2$–C(=O)–CH$_2$CH$_3$ diethyl ketone (3-pentanone)

74. a. CH$_3$—CH$_2$—C(=O)—C$_6$H$_5$

b. CH$_3$–CH$_2$–C(=O)–CH$_2$–CH$_3$

c. CH$_3$–CH$_2$–C(=O)–CH$_2$–CH$_3$ (same molecule as above)

d. O=C(H)–CH(Cl)–CH$_3$

e. O=C(H)–CH(CH$_3$)–CH(CH$_3$)–CH$_2$–CH$_3$

76. Carboxylic acids are typically *weak* acids.

CH$_3$–CH$_2$–COOH(aq) ⇌ H$^+$(aq) + CH$_3$–CH$_2$–COO$^-$(aq)

78. a. CH$_2$–CH$_2$–COOH

b. CH$_3$–CH$_2$–C(=O)–CH$_3$

c. no reaction (tertiary alcohols are not oxidized in this manner)

80. Acetylsalicylic acid is synthesized from salicylic acid (behaving as an alcohol through its –OH group) and acetic acid.

[salicylic acid (C$_6$H$_4$(OH)(COOH)) + HO–C(=O)–CH$_3$ → acetylsalicylic acid (C$_6$H$_4$(COOH)(O–C(=O)–CH$_3$)) + H$_2$O]

82. a.

$$CH_3-CH_2-CH_2-\overset{\overset{O}{\|}}{C}-O-CH_2-CH_3$$

b.

[Structure: benzene ring with COOH, OH (ortho), and Cl (para to COOH)]

c.

$$CH_3-CH_2-CH_2-CH_2-\overset{\overset{O}{\|}}{C}-O-CH_3$$

d.

[Structure: benzene ring with $-\overset{\overset{O}{\|}}{C}-O-CH_2-CH_2-CH_3$]

e.

$$CH_3-\underset{\underset{}{}}{\overset{\overset{CH_3}{|}}{CH}}-\underset{\underset{}{}}{\overset{\overset{CH_3}{|}}{CH}}-C\overset{\nearrow O}{\underset{\searrow OH}{}}$$

84. In addition polymerization, the monomer units simply add together to form the polymer, with no other products. Polyethylene and polytetrafluoroethylene (Teflon) are common examples.

86. A polyester is formed from the reaction of a dialcohol (two –OH groups) with a diacid (two –COOH groups). One –OH group of the alcohol forms an *ester linkage* with one of the –COOH groups of the acid. Since the resulting dimer still possesses an –OH and a –COOH group, the dimer can undergo further esterification reactions. Dacron is a common polyester.

88.

nylon

$$-\left(-N(H)-(CH_2)_6-N(H)-C(=O)-(CH_2)_4-C(=O)-\right)-$$

dacron

$$-\left(-O-CH_2-CH_2-O-C(=O)-\underset{}{\bigcirc}-C(=O)-\right)-$$

90. saturated

92. straight-chain or normal

94. -ane

96. number

98. anti-knocking

100. substitution

102. hydrogenation

104. functional group

106. carbon monoxide

108. carbonyl

110. carboxyl

112. addition

114. The carbon skeletons are

```
C-C-C-C-C           C-C-C-C
    |                   |
    C                   C
    |
  C-C-C
    |
    C
```

116. a. 2-chlorobutane
 b. 1,2-dibromoethane
 c. triiodomethane (common name: iodoform)
 d. 2,3,4-trichloropentane
 e. 2,2-dichloro-4-isopropylheptane

118. a. $\mathrm{CH_3-CH_2-\underset{\underset{Cl}{|}}{C}=CH-CH_2-CH_3}$

b. $\mathrm{\underset{\underset{Br}{|}}{CH_2}-\underset{\underset{Br}{|}}{CH_2}}$

c. $\mathrm{\underset{\underset{Cl}{|}}{\overset{\overset{Cl}{|}}{CH}}-\underset{\underset{Cl}{|}}{\overset{\overset{Cl}{|}}{CH}}}$

d. $\mathrm{CH\equiv C-\underset{\underset{Br}{|}}{CH}-CH_2-CH_2-CH_3}$

e. $\mathrm{\underset{\underset{Cl}{|}}{CH_2}-\underset{\underset{Cl}{|}}{\overset{\overset{CH_3}{|}}{C}}---\underset{\underset{CH_3}{|}}{\overset{\overset{CH_3}{|}}{C}}---\underset{\underset{CH_3}{|}}{CH}-CH_2-CH_2-CH_2-CH_3}$

120. primary $\mathrm{CH_3-CH_2-CH_2-CH_2-CH_2-CH_2-OH}$

secondary $\mathrm{CH_3-CH_2-CH_2-CH_2-\underset{\underset{OH}{|}}{CH}-CH_3}$

tertiary $\mathrm{CH_3-CH_2-\underset{\underset{OH}{|}}{\overset{\overset{CH_3}{|}}{C}}-CH_2-CH_3}$

122. $\mathrm{\underset{\underset{OH}{|}}{CH_2}-\underset{\underset{OH}{|}}{CH}-\underset{\underset{OH}{|}}{CH}-\underset{\underset{OH}{|}}{CH}-\underset{\underset{OH}{|}}{CH}-\underset{\underset{H}{|}}{C}=O}$

124. a.

   (diphenyl ketone structure: two phenyl rings attached to C=O)

b. $\mathrm{CH_3-\underset{\underset{OH}{|}}{\overset{\overset{CH_3}{|}}{C}}-CH_3}$

258   Chapter 20   Organic Chemistry

c.

$$\begin{array}{c} CH_3 \ \ O \\ | \ \ \ \ \ || \\ CH-C-CH_3 \\ | \\ CH_3 \end{array}$$

d.

$$\begin{array}{c} CH_2-CH_2 \\ | \ \ \ \ \ \ | \\ OH \ \ \ OH \end{array}$$

e.

$$CH_3-CH_2-C\begin{array}{c} {}^{\nearrow O} \\ {}_{\searrow O-CH_3} \end{array}$$

126.

$$\begin{array}{c} HO-C=O \\ | \\ CH_3-CH-N-(H+HO)-C=O \rightarrow \\ | \ | \\ H \ \ \ \ \ \ \ \ \ \ \ \ \ CH_2-NH_2 \end{array}$$

$$\begin{array}{c} HO-C=O \\ | \\ CH_3-CH-N-C=O + H_2O \\ | \ \ \ \ \ \ \ | \\ H \ \ \ CH_2-NH_2 \end{array}$$

One end has —NH$_2$, which can react with the —COOH end of another of these dipeptides.

128. a.

$$\begin{array}{c} H \ H \ H \ H \ H \ H \ H \ H \\ | \ | \ | \ | \ | \ | \ | \ | \\ H-C-C-C-C-C-C-C-C-H \\ | \ | \ | \ | \ | \ | \ | \ | \\ H \ H \ H \ H \ H \ H \ H \ H \end{array}$$

octane CH$_3$CH$_2$CH$_2$CH$_2$CH$_2$CH$_2$CH$_2$CH$_3$

b.

$$\begin{array}{c} H \ H \ H \ H \ H \ H \\ | \ | \ | \ | \ | \ | \\ H-C-C-C-C-C-C-H \\ | \ | \ | \ | \ | \ | \\ H \ H \ H \ H \ H \ H \end{array}$$

hexane CH$_3$CH$_2$CH$_2$CH$_2$CH$_2$CH$_3$

c.
```
     H H H H
     | | | |
   H-C-C-C-C-H
     | | | |
     H H H H
```
butane CH₃CH₂CH₂CH₃

d.
```
     H H H H H
     | | | | |
   H-C-C-C-C-C-H
     | | | | |
     H H H H H
```
pentane CH₃CH₂CH₂CH₂CH₃

130.
a. 2,3-dimethylbutane
b. 3,3-diethylpentane
c. 2,3,3-trimethylhexane
d. 2,3,4,5,6-pentamethylheptane

132.
a. CH₃Cl(g)
b. H₂(g)
c. HCl(g)

134.
CH≡C–CH₂–CH₂–CH₂–CH₂–CH₂–CH₃   1-octyne

CH₃–C≡C–CH₂–CH₂–CH₂–CH₂–CH₃   2-octyne

CH₃–CH₂–C≡C–CH₂–CH₂–CH₂–CH₃   3-octyne

CH₃–CH₂–CH₂–C≡C–CH₂–CH₂–CH₃   4-octyne

136.
a. carboxylic acid
b. ketone
c. ester
d. alcohol (phenol)

138.
a. 3-methylpentanal

$$CH_3-CH_2-\underset{\underset{CH_3}{|}}{CH}-CH_2-\underset{\underset{H}{|}}{C}=O$$

b. 3-methyl-2-pentanone

$$CH_3-CH_2-\underset{\underset{CH_3}{|}}{CH}-\underset{\underset{O}{\|}}{C}-CH_3$$

c. methyl phenyl ketone

$$CH_3-\underset{\underset{O}{\|}}{C}-\underset{\bigcirc}{}$$

d. 2-hydroxybutanal

$$CH_3-CH_2-\underset{\underset{OH}{|}}{CH}-\underset{\underset{H}{|}}{C}=O$$

e. propanal

$$CH_3-CH_2-\underset{\underset{H}{|}}{C}=O$$

140. a. $$CH_3-\underset{\underset{CH_3}{|}}{CH}-CH_2-COOH$$

b.

benzene ring with C(=O)-OH and Cl substituents

c. $CH_3-CH_2-CH_2-CH_2-CH_2-COOH$

d. CH_3-COOH

Chapter 21   Biochemistry

2. Proteins represent biopolymers of α-amino acids used for many purposes in the human body (structure, enzymes, antibodies, etc.). Proteins make up about 15% of the body by mass.

4. Fibrous proteins provide the structural material of many tissues in the body, and are the chief constituents of hair, cartilage, and muscles. Fibrous proteins consist of lengthwise bundles of polypeptide chains (a fiber). Globular proteins have their polypeptide chains folded into a basically spherical shape and tend to be found in the bloodstream, where they transport and store various needed substances, act as antibodies to fight infections, act as enzymes to catalyze cellular processes, participate in the body's various regulatory systems, and so on.

6. The structures of the amino acids are given in detail in Figure 21.2. Generally a side chain is nonpolar if it is mostly hydrocarbon in nature (e.g., alanine, in which the side chain is a methyl group). Side chains are polar if they contain the hydroxyl group (–OH), the sulfhydryl group (–SH), or a second amino (–NH$_2$) or carboxyl (–COOH) group.

8. Figure 21.2 shows the amino acids separated into the hydrophilic and hydrophobic subgroupings. Notice that the hydrophobic amino acids contain R groups which are basically hydrocarbon (or substituted hydrocarbon) in nature. This makes these R groups nonpolar and unlikely to interact with very polar water molecules. Notice that the hydrophilic amino acids all contain a very polar functional group (e.g., –OH, –SH, –C=O, etc.) which enable the amino acid R groups to interact with water. For a protein in an aqueous medium, the hydrophilic R groups will orient themselves towards the aqueous medium, while the hydrophobic R groups will turn away from the aqueous medium, causing a profound effect on the protein's shape.

10. There is a very large number of polypeptides possible. From this exercise, you can certainly appreciate the complexities and intricacies of protein structures.

ala-phe-cys-lys	*cys-ala-phe-lys*
ala-cys-lys-phe	*cys-ala-lys-phe*
ala-lys-cys-phe	*lys-ala-cys-phe*
ala-phe-lys-cys	*lys-ala-phe-cys*
ala-lys-phe-cys	*lys-phe-ala-cys*
ala-cys-phe-lys	*lys-phe-cys-ala*
phe-ala-lys-cys	*lys-cys-phe-ala*
phe-ala-cys-lys	*lys-cys-ala-phe*
phe-cys-lys-ala	*cys-phe-ala-lys*
cys-phe-lys-ala	*cys-lys-phe-ala*
cys-lys-ala-phe	*phe-cys-ala-lys*
phe-lys-cys-ala	*phe-lys-ala-cys*

262 Chapter 21   Biochemistry

12. phe-ala-gly

$$\underset{\text{N terminal}}{H_2N-}\overset{\overset{\displaystyle CH_2-C_6H_5}{|}}{\underset{|}{C}}-\overset{O}{\overset{\|}{C}}-N-\overset{\overset{\displaystyle CH_3}{|}}{\underset{|}{C}}-\overset{O}{\overset{\|}{C}}-N-\overset{H}{\underset{|}{C}}-COOH \;\;\text{C terminal}$$

phe-gly-ala

$$H_2N-\overset{CH_2-C_6H_5}{\underset{|}{C}}-\overset{O}{\overset{\|}{C}}-N-\overset{H}{\underset{|}{C}}-\overset{O}{\overset{\|}{C}}-N-\overset{CH_3}{\underset{|}{C}}-COOH$$

ala-phe-gly

$$H_2N-\overset{CH_3}{\underset{|}{C}}-\overset{O}{\overset{\|}{C}}-N-\overset{CH_2-C_6H_5}{\underset{|}{C}}-\overset{O}{\overset{\|}{C}}-N-\overset{H}{\underset{|}{C}}-COOH$$

ala-gly-phe

$$H_2N-\overset{CH_3}{\underset{|}{C}}-\overset{O}{\overset{\|}{C}}-N-\overset{H}{\underset{|}{C}}-\overset{O}{\overset{\|}{C}}-N-\overset{CH_2-C_6H_5}{\underset{|}{C}}-COOH$$

gly-phe-ala

$$H_2N-\overset{H}{\underset{|}{C}}-\overset{O}{\overset{\|}{C}}-N-\overset{CH_2-C_6H_5}{\underset{|}{C}}-\overset{O}{\overset{\|}{C}}-N-\overset{CH_3}{\underset{|}{C}}-COOH$$

gly-ala-phe

$$H_2N-\overset{H}{\underset{|}{C}}-\overset{O}{\overset{\|}{C}}-N-\overset{H}{\underset{|}{C}}-\overset{O}{\overset{\|}{C}}-N-\overset{CH_2-C_6H_5}{\underset{|}{C}}-COOH$$

14. Long, thin, resilient proteins, such as hair, typically contain elongated, elastic alpha-helical protein molecules. Other proteins, such as silk, which in bulk form sheets or plates, typically contain protein molecules having the beta pleated sheet secondary structure. Proteins which do not have a structural function in the body, such as hemoglobin, typically have a globular structure.

16. The secondary structure, in general, describes the arrangement of the long polypeptide chain of the protein. In the alpha-helical secondary structure, the chain forms a coil or spiral, which gives proteins consisting of such structures an elasticity or resilience. Such proteins are found in wool, hair, and tendons.

18. Cysteine, an amino acid containing the sulfhydryl (-SH) group in its side chain, is capable of forming disulfide linkages (-S-S-) with other cysteine molecules in the same polypeptide chain. If such a disulfide linkage is formed, this effectively ties together two portions of the polypeptide, producing a kink or knot in the chain, which leads in part to the protein's overall 3-dimensional shape (tertiary structure). Cysteine, and the disulfide linkages it forms, is responsible for the curling of hair (whether naturally or by a permanent wave).

20. Collagen has an alpha-helical secondary structure. Collagen's function in the body is as the raw material of which tendons are constructed. The long, springy structure of the alpha-helix is responsible for collagen's strength and elasticity.

22. Antibodies are special proteins which are synthesized in the body in response to foreign substances such as bacteria or viruses. Although there are usually specific antibodies for specific invaders, the antibody interferon offers general protection against invasion by viruses.

24. In a permanent wave, cross-linkages between adjacent polypeptide chains of the protein are broken chemically, and then reformed chemically in a new location. The primary cross-linkage involved is a disulfide linkage between cysteine units in the polypeptide chains. It is primarily the tertiary structure of the hair protein which is affected by a permanent wave, although if the waving lotion is left on the hair too long, the secondary structure can also be affected (making the hair very "frizzy").

26. The molecule acted upon by an enzyme is referred to as the enzyme's substrate. When we say that an enzyme is specific for a particular substrate, we mean that the enzyme will catalyze the reactions of that molecule and that molecule only.

28. Figure 21.11 illustrates the lock and key model clearly. The lock-and-key model for enzyme action postulates that the structures of the enzyme and its substrate are complementary, such that the active site of the enzyme and the portion of the substrate molecule to be acted upon can fit together very closely. The structures of these portions of the molecules are unique to the particular enzyme-substrate pair, and they fit together much like a particular key is necessary to work in a given lock.

30. Sugars contain an aldehyde or ketone functional group (carbonyl group), as well as several (OH groups (hydroxyl group).

32.   a.   glucose

```
      CHO
       |
  H — C — OH
       |
 HO — C — H
       |
  H — C — OH
       |
  H — C — OH
       |
      CH₂OH
```

   b.   ribose

```
      CHO
       |
  H — C — OH
       |
  H — C — OH
       |
  H — C — OH
       |
      CH₂OH
```

   c.   ribulose

```
      CH₂OH
       |
      C = O
       |
  H — C — OH
       |
  H — C — OH
       |
      CH₂OH
```

   d.   galactose

```
      CHO
       |
  H — C — OH
       |
 HO — C — H
       |
 HO — C — H
       |
  H — C — OH
       |
      CH₂OH
```

34. A polysaccharide is a large polymeric substance, containing as its building blocks repeating simple sugar (monosaccharide) monomer units. Both starch and cellulose consist of long chains of bonded *glucose* molecules: the differences in properties between starch and cellulose are derived from the different manner in which the glucose units are attached to one another to form the polysaccharide chains.

36. ribose (aldopentose); arabinose (aldopentose); ribulose (ketopentose); glucose (aldohexose); mannose (aldohexose); galactose (aldohexose); fructose (ketohexose).

38. A nucleotide consists of three components: a five-carbon sugar, a nitrogen-containing organic base, and a phosphate group. The phosphate and nitrogen base are each bonded to respective sites on the sugar, but are not bonded to each other.

40. Uracil (RNA only); cytosine (DNA, RNA); thymine (DNA only); adenine (DNA, RNA); guanine (DNA, RNA)

42. When the two strands of a DNA molecule are compared, it is found that a given base in one strand is always found paired with a particular base in the other strand. Because of the shapes and side atoms along the rings of the nitrogen bases, only certain pairs are able to approach and hydrogen-bond with each other in the double helix. Adenine is always found paired with thymine; cytosine is always found paired with guanine. When a DNA helix unwinds for replication during cell division, only the appropriate complementary bases are able to approach and bond to the nitrogen bases of each strand. For example, for a guanine-cytosine pair in the original DNA, when the two strands separate, only a new cytosine molecule can approach and bond to the original guanine, and only a new guanine molecule can approach and bond to the original cytosine.

44. Messenger RNA molecules are synthesized to be complementary to a portion (gene) of the DNA molecule in the cell, and serve as the template or pattern upon which a protein will be constructed (a particular group of nitrogen bases on *m*-RNA is able to accommodate and specify a particular amino acid in a particular location in the protein). Transfer RNA molecules are much smaller than *m*-RNA, and their structure accommodates only a single specific amino acid molecule: transfer RNA molecules "find" their specific amino acid in the cellular fluids, and bring it to *m*-RNA where it is added to the protein molecule being synthesized.

46. A triglyceride typically consists of a glycerol backbone, to which three separate fatty acid molecules are attached by ester linkages

$$\begin{array}{l} CH_2-O-\overset{\overset{O}{\|}}{C}-R \\ CH-O-\overset{\overset{O}{\|}}{C}-R' \\ CH_2-O-\overset{\overset{O}{\|}}{C}-R'' \end{array}$$

48. Saponification is the production of a *soap* by treatment of a triglyceride with a strong base such as NaOH.

triglyceride + 3NaOH → glycerol + 3Na⁺soap⁻

$$\begin{array}{l}CH_2-O-\overset{O}{\overset{\|}{C}}-R \\ CH-O-\overset{O}{\overset{\|}{C}}-R' \\ CH_2-O-\overset{O}{\overset{\|}{C}}-R''\end{array} + 3NaOH \rightarrow \begin{array}{l}CH_2-OH \\ CH-OH \\ CH_2-OH\end{array} + \begin{array}{l}RCOONa \\ R'COONa \\ R''COONa\end{array}$$

50. Soaps have both a nonpolar nature (due to the long chain of the fatty acid) and an ionic nature (due to the charge on the carboxyl group). In water, soap anions form aggregates called micelles, in which the water-repelling hydrocarbon chains are oriented towards the interior of the aggregate, with the ionic, water-attracting carboxyl groups oriented towards the outside. Most dirt has a greasy nature. A soap micelle is able to interact with a grease molecule, pulling the grease molecule into the hydrocarbon interior of the micelle. When the clothing is rinsed, the micelle containing the grease is washed away. See Figures 21.22 and 21.23.

52. Cholesterol is the naturally occurring steroid from which the body synthesizes other needed steroids. Since cholesterol is insoluble in water, it is thought that having too large a concentration of this substance in the bloodstream may lead to its deposition and build up on the walls of blood vessels, causing their eventual blockage.

54. Phospholipids have a similar structure to triglycerides, except that one of the fatty acids attached to the glycerol backbone is replaced by a phosphate group (phospholipids are sometimes referred to as phosphodiglycerides). An important phospholipid is lecithin, which behaves very strongly as an emulsifying agent (allowing polar and nonpolar substances to mix). Lecithin is found in high concentration in egg yolks, and is being used in baking in place of whole eggs: the emulsifying properties of the egg yolk are due to the lecithin, and using lecithin itself avoids the cholesterol problem associated with whole eggs.

56. i
58. m
60. u
62. f
64. g
66. r

68. p

70. o

72. b

74. d

76. a

78. nucleotides

80. ester

82. adenine, cytosine

84. transfer, messenger

86. insoluble, soluble

88. unsaturated, saturated

90. ionic, nonpolar

92. waxes

94. progesterone

96. The primary structure of a protein refers to the specific identity and ordering of amino acids in a protein's polypeptide chain. The primary structure is sometimes referred to as the protein's amino acid *sequence*.

98. tendons, bone (with mineral constituents), skin, cartilage, hair, fingernails.

100. Proteins contain both acidic (−COOH) and basic (−NH$_2$) groups in their side chains, which can neutralize both acids and bases.

102. pentoses (5 carbons); hexoses (6 carbons); trioses (3 carbons)

104. In a strand of DNA, the phosphate group and the sugar molecule of adjacent nucleotides become bound to each other. The chain-portion of the DNA molecule, therefore, consists of alternating phosphate groups and sugar molecules. The nitrogen bases are found sticking out from the side of this phosphate-sugar chain, being bonded to the sugar molecules.

106. The primary human bile acid is cholic acid, which helps to emulsify fats in the intestinal tract. In order to be digested by enzymes for absorption into the bloodstream, large clumps of fat must be dispersed into fine droplets in the liquid of the small intestine. Cholic acid basically acts as a detergent.